BIOINFORMATICS PRACTICAL MANUAL

Author Message:

All inquiries should be emailed to ***mohammed@biocurationssl.com*** ***,biocurationssl@gmail.com***

PUBLISHING FOR ONE WORLD

Biocuration Labs
#15/7,7th cross , Someshwaranagar ,
Jayanagar 1st block,Bangalore-560011
Ph: 08026560388 , M:08553794025
Visit us at **www.biocurationssl.com**

PREFACE

This book has originated from Practical class on bioinformatics that are offered to students of Bioinformatics, Bangalore University of Bangalore. The idea to write a book on Bioinformatics was born during the preparations of these practical where I realized that it is extremely difficult to achieve an overview of the area of Bioinformatics and to follow the progress of this field. This is the first book in 2014 and was written in English .Bioinformatics is a major topic in modern medical, Life science and pharmacological research and is of central importance in the computational biology science. Accordingly, The enormous increase in data on Bioinformatics has led me to leave out the practical on bioinformatics and. This topic has since evolved into a huge research area of its own that could not be considered adequately within this book. My knowledge of Bioinformatics practical has exploded in the past 5 years,Bioinformatics could be treated here with the same thoroughness. It is the aim of the present book to describe the Bioinformatics practical approach for life science students.

I am grateful to all people who have encouraged me to continue with the book and Who have supported me with many helpful comments and corrections. In first place I Want to thank my colleague Prof Mohammed Rukunnudin Ghalib . I am want to thankful to students who give me suggestion to improve for next edition as per requirement.

CONTENT

1. ACCESSING BIBLIOGRAPHIC DATABASE FROM NCBI

- Open internet explorer.
- Go to http://www.ncbi.nlm.nih.gov.
- Select pub med from NCBI home page --→ select All database drop down box and in search area text box type query search such as pubmed id or name of organism for article collection.
- Click go button for displaying the result sheet.
- The result sheet with bibliographic information will be displayed.
- Click on the hyperlinked on article name of your interest.
- The result page will be displayed.

Save the file by save as or copy the article {bibliography to the note pad} and save

Bibilograpic Database

Result:

The given Organism **Solanum Surrattense** PUBMED ID :**23427398**

Bibilographic Database is Found.

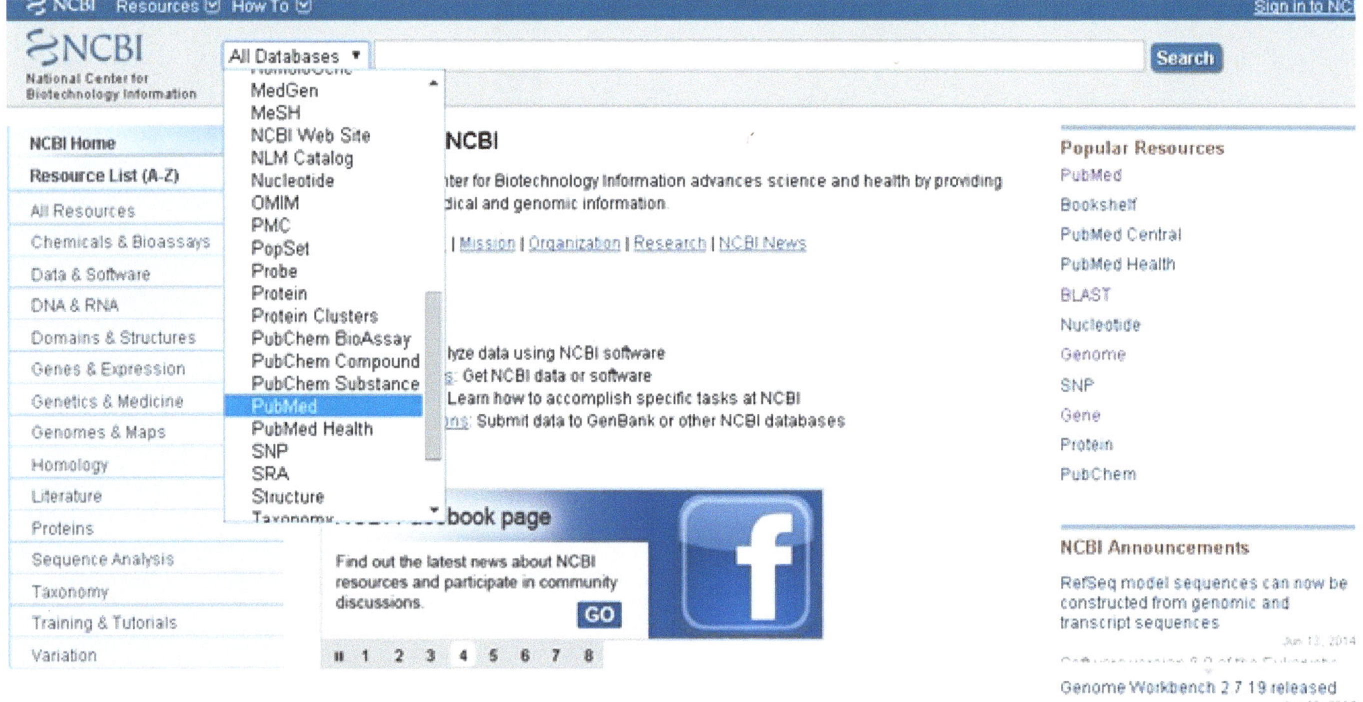

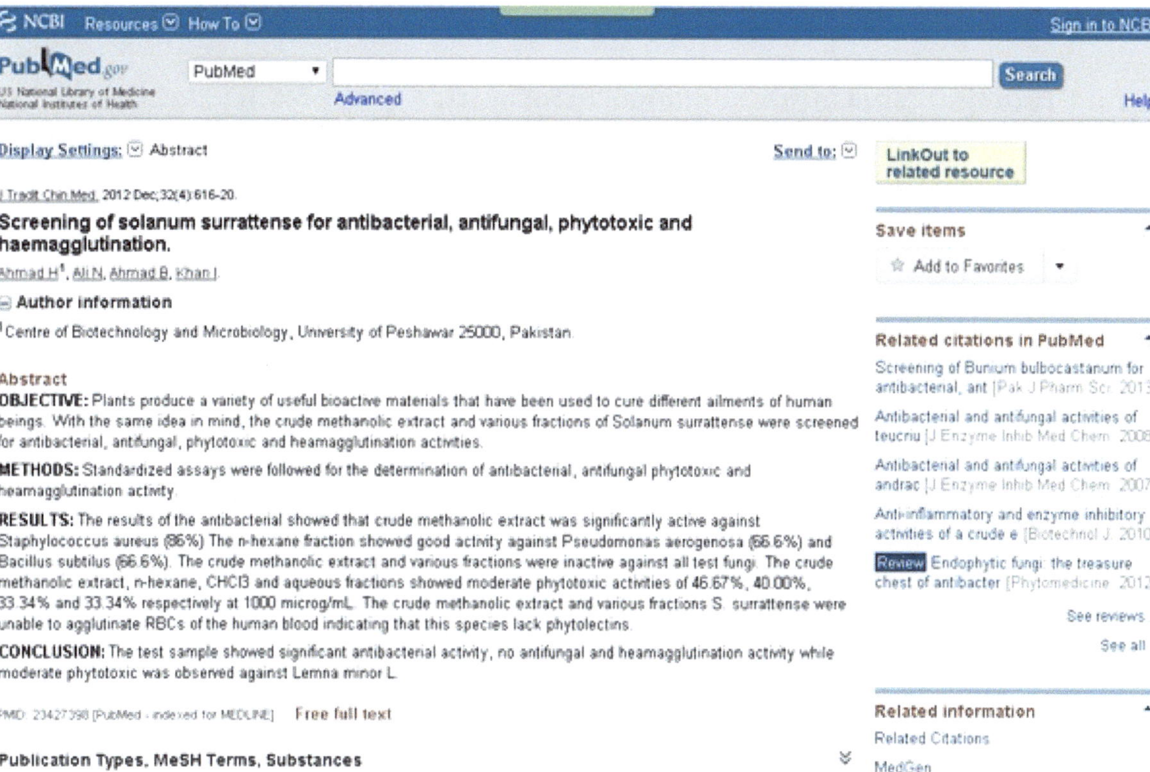
NCBI Resources How To Sign in to NCBI
PubMed.gov US National Library of Medicine National Institutes of Health
PubMed Search
Advanced Help

Display Settings: Abstract Send to:

J Tradit Chin Med. 2012 Dec;32(4):616-20.

Screening of solanum surrattense for antibacterial, antifungal, phytotoxic and haemagglutination.

Ahmad H[1], Ali N, Ahmad B, Khan I.

Author information

[1]Centre of Biotechnology and Microbiology, University of Peshawar 25000, Pakistan.

Abstract

OBJECTIVE: Plants produce a variety of useful bioactive materials that have been used to cure different ailments of human beings. With the same idea in mind, the crude methanolic extract and various fractions of Solanum surrattense were screened for antibacterial, antifungal, phytotoxic and heamagglutination activities.

METHODS: Standardized assays were followed for the determination of antibacterial, antifungal phytotoxic and heamagglutination activity.

RESULTS: The results of the antibacterial showed that crude methanolic extract was significantly active against Staphylococcus aureus (86%) The n-hexane fraction showed good activity against Pseudomonas aerogenosa (66.6%) and Bacillus subtilus (66.6%). The crude methanolic extract and various fractions were inactive against all test fungi. The crude methanolic extract, n-hexane, CHCl3 and aqueous fractions showed moderate phytotoxic activities of 46.67%, 40.00%, 33.34% and 33.34% respectively at 1000 microg/mL. The crude methanolic extract and various fractions S. surrattense were unable to agglutinate RBCs of the human blood indicating that this species lack phytolectins.

CONCLUSION: The test sample showed significant antibacterial activity, no antifungal and heamagglutination activity while moderate phytotoxic was observed against Lemna minor L.

PMID: 23427398 [PubMed - indexed for MEDLINE] Free full text

Publication Types, MeSH Terms, Substances

LinkOut to related resource

Save items
Add to Favorites

Related citations in PubMed
Screening of Bunium bulbocastanum for antibacterial, ant [Pak J Pharm Sci. 2013]
Antibacterial and antifungal activities of teucriu [J Enzyme Inhib Med Chem. 2008]
Antibacterial and antifungal activities of andrac [J Enzyme Inhib Med Chem. 2007]
Anti-inflammatory and enzyme inhibitory activities of a crude e [Biotechnol J. 2010]
Review Endophytic fungi: the treasure chest of antibacter [Phytomedicine. 2012]
See reviews
See all

Related information
Related Citations
MedGen

2. Retrieving Protein / Nucleotide sequence information from NCBI Database

Retrieving Nucleotide sequence

- Open any Internet Browser
- Go to the link http://www.ncbi.nlm.nih.gov
- On the top of NCBI home page drag down All database and Select Protein.
- In the search text box type key word of query such as Accession number (or) disease name (or) organism name to retrieve the sequence.
- Click on Search button this will display list of available result match to the given keyword.
- Select the check box in the result page which you want to retrieve the sequence.
- Go to display setting and select FASTA format .Because most of the bioinformatics tools accept FASTA file format.
- The result will display in FASTA format
- Save the file by save as in the top right corner by clicking button send file
    - Or Copy the file text from '>' (greater symbol) up to the End of the sequence text and save in notepad with file name and extension as **.FASTA .**

Sequence Retrieving

Result : The given Organism Brassica oleracea ,GenBank: BH947338.1, gi|23427398| Nucleotide sequence is retrieved in FASTA format

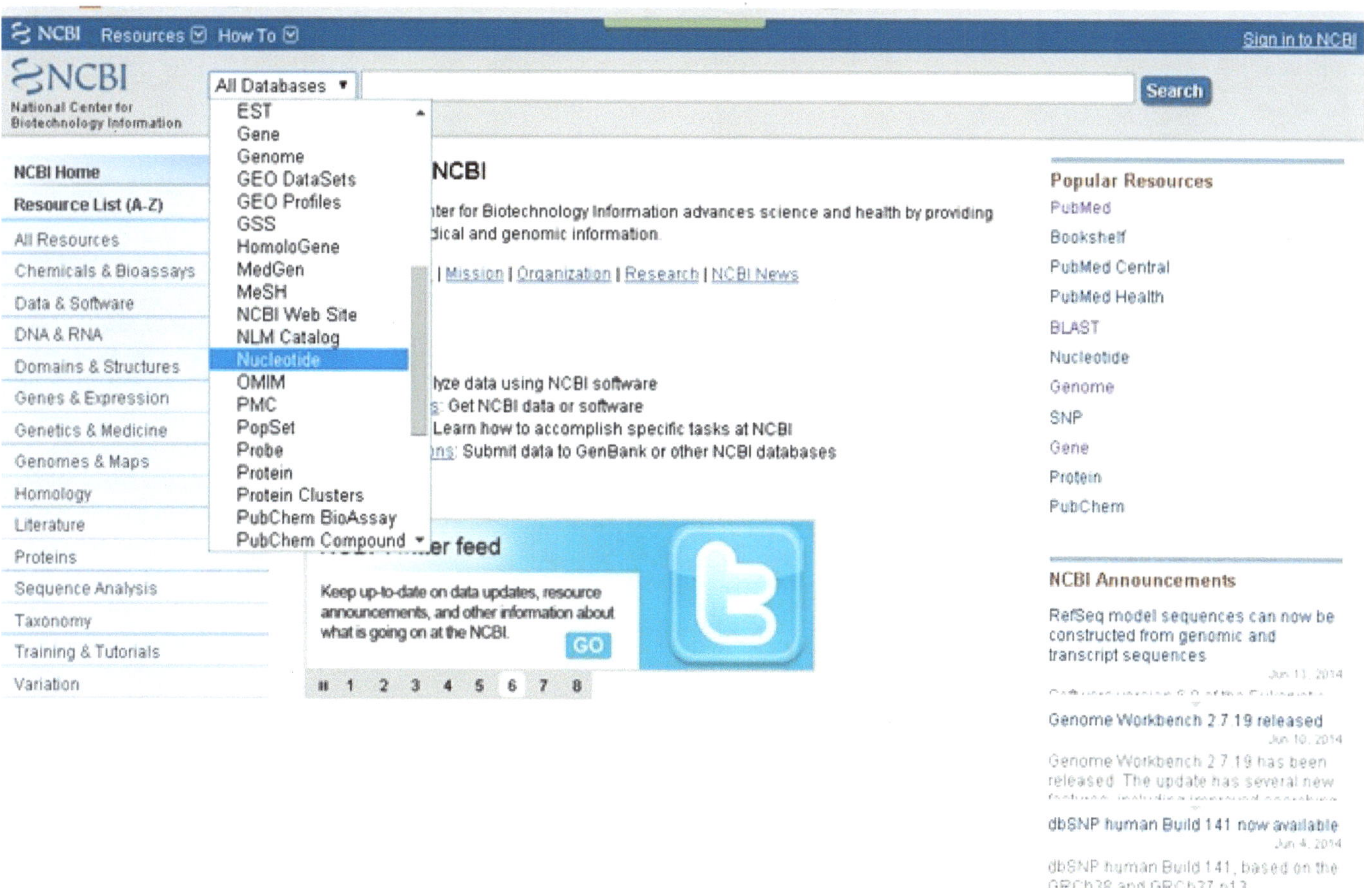

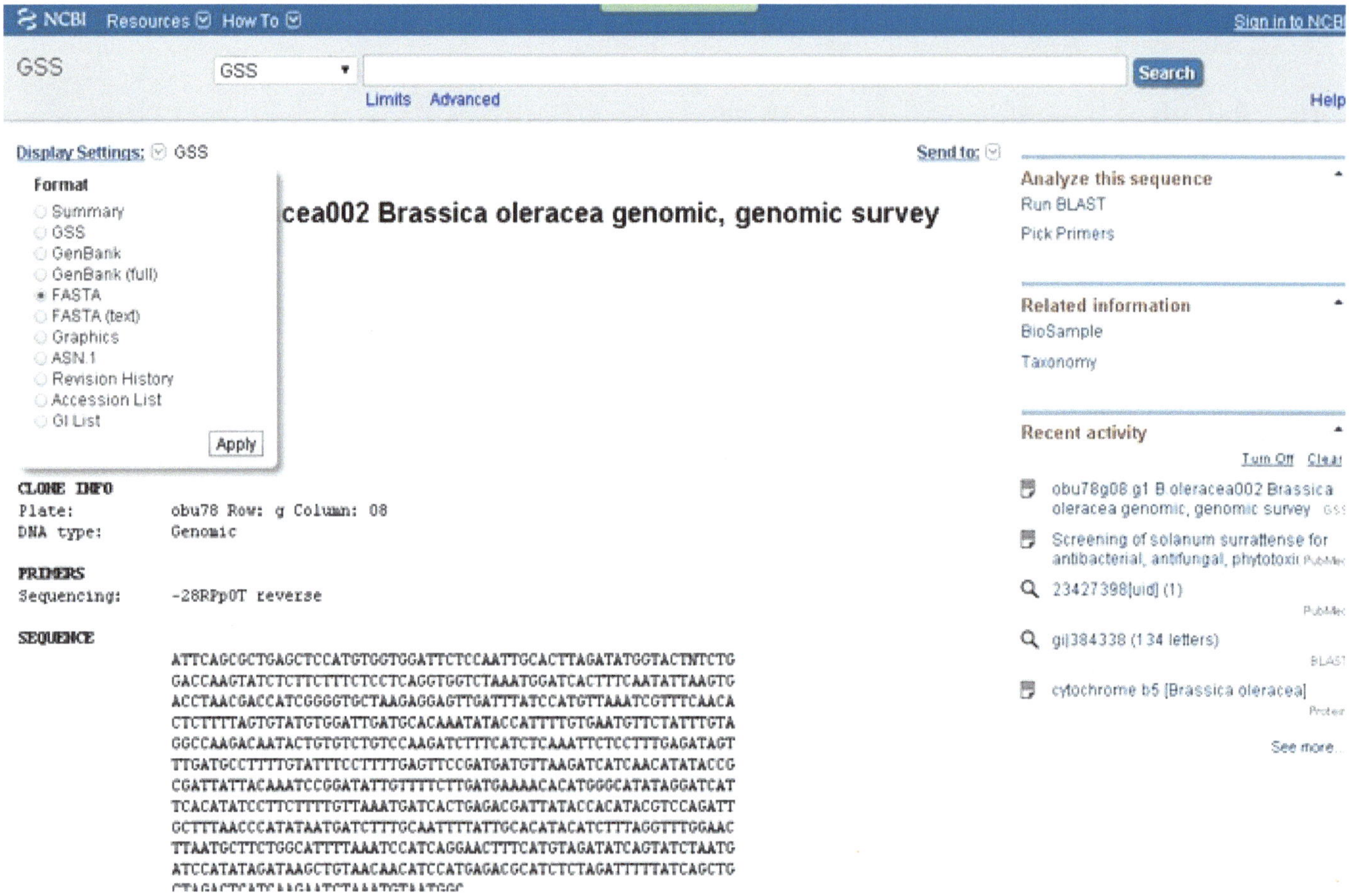

NCBI Resources How To Sign in to NCBI

GSS | GSS | Search | Limits | Advanced | Help

Display Settings: GSS | Send to:

Format
- Summary
- GSS
- GenBank
- GenBank (full)
- FASTA
- FASTA (text)
- Graphics
- ASN.1
- Revision History
- Accession List
- GI List

Apply

...cea002 Brassica oleracea genomic, genomic survey

Analyze this sequence
Run BLAST
Pick Primers

Related information
BioSample
Taxonomy

Recent activity
Turn Off Clear
obu78g08.g1 B.oleracea002 Brassica oleracea genomic, genomic survey GSS
Screening of solanum surrattense for antibacterial, antifungal, phytotoxi PubMed
23427398[uid] (1) PubMed
gi|384338 (134 letters) BLAST
cytochrome b5 [Brassica oleracea] Protein
See more...

```
CLONE INFO
Plate:        obu78 Row: g Column: 08
DNA type:     Genomic

PRIMERS
Sequencing:   -28RPp0T reverse

SEQUENCE
              ATTCAGCGCTGAGCTCCATGTGGTGGATTCTCCAATTGCACTTAGATATGGTACTNTCTG
              GACCAAGTATCTCTTCTTTCTCCTCAGGTGGTCTAAATGGATCACTTTCAATATTAAGTG
              ACCTAACGACCATCGGGGTGCTAAGAGGAGTTGATTTATCCATGTTAAATCGTTTCAACA
              CTCTTTTAGTGTATGTGGATTGATGCACAAATATACCATTTTGTGAATGTTCTATTTGTA
              GGCCAAGACAATACTGTGTCTGTCCAAGATCTTTCATCTCAAATTCTCCTTTGAGATAGT
              TTGATGCCTTTTGTATTTCCTTTTGAGTTCCGATGATGTTAAGATCATCAACATATACCG
              CGATTATTACAAATCCGGATATTGTTTTCTTGATGAAAACACATGGGCATATAGGATCAT
              TCACATATCCTTCTTTTGTTAAATGATCACTGAGACGATTATACCACATACGTCCAGATT
              GCTTTAACCCATATAATGATCTTTGCAATTTTATTGCACATACATCTTTAGGTTTGGAAC
              TTAATGCTTCTGGCATTTTAAATCCATCAGGAACTTTCATGTAGATATCAGTATCTAATG
              ATCCATATAGATAAGCTGTAACAACATCCATGAGACGCATCTCTAGATTTTTATCAGCTG
```

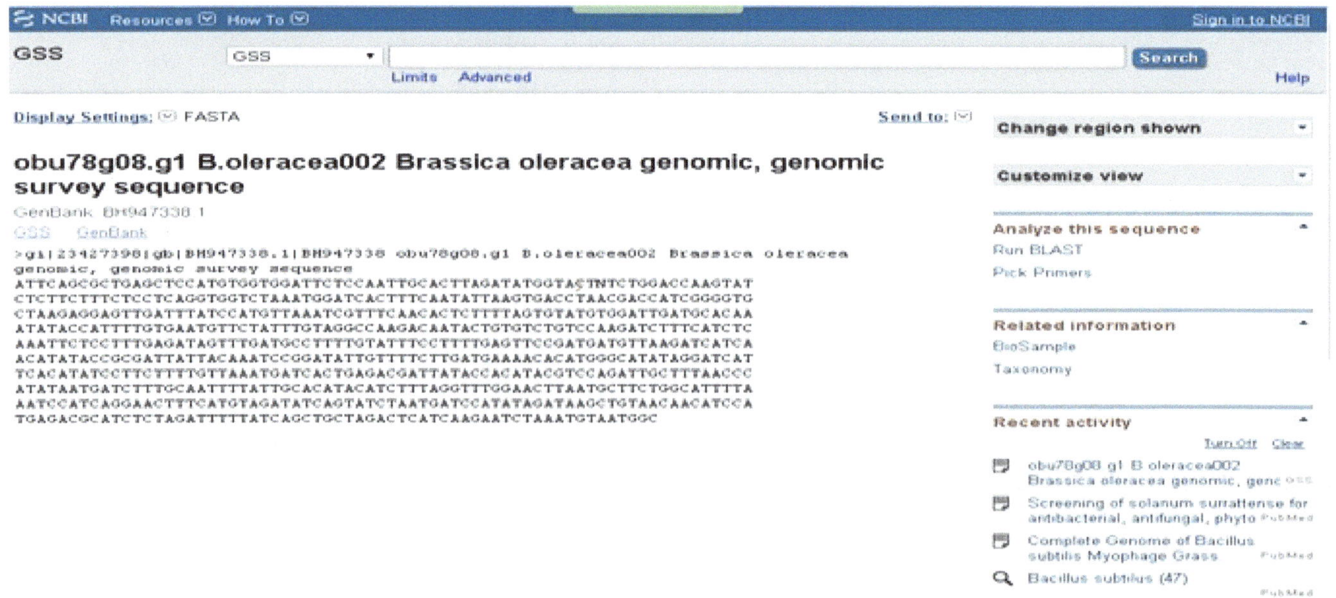

Retrieving Protein sequence

- Open any Internet Browser
- Go to the link http://www.ncbi.nlm.nih.gov
- On the top of NCBI home page drag down All database and Select Protein.
- In the search text box type key word of query such as Accession number (or) protein name (or) disease name (or) organism name to retrieve the sequence.
- Click on Search button this will display list of available result match to the given keyword.
- Select the check box in the result page which you want to retrieve the sequence.
- Go to display setting and select FASTA format .Because most of the bioinformatics tools accept FASTA file format.
- The result will display in FASTA format
- Save the file by save as in the top right corner by clicking button send file
- Or Copy the file text from '>' (greater symbol) up to the End of the sequence text and save in notepad with file name and extension as **.FASTA .**

Result : The given Organism Brassica oleracea , Accession : 1905426A gi|384338|Protein sequence is retrieved in FASTA format

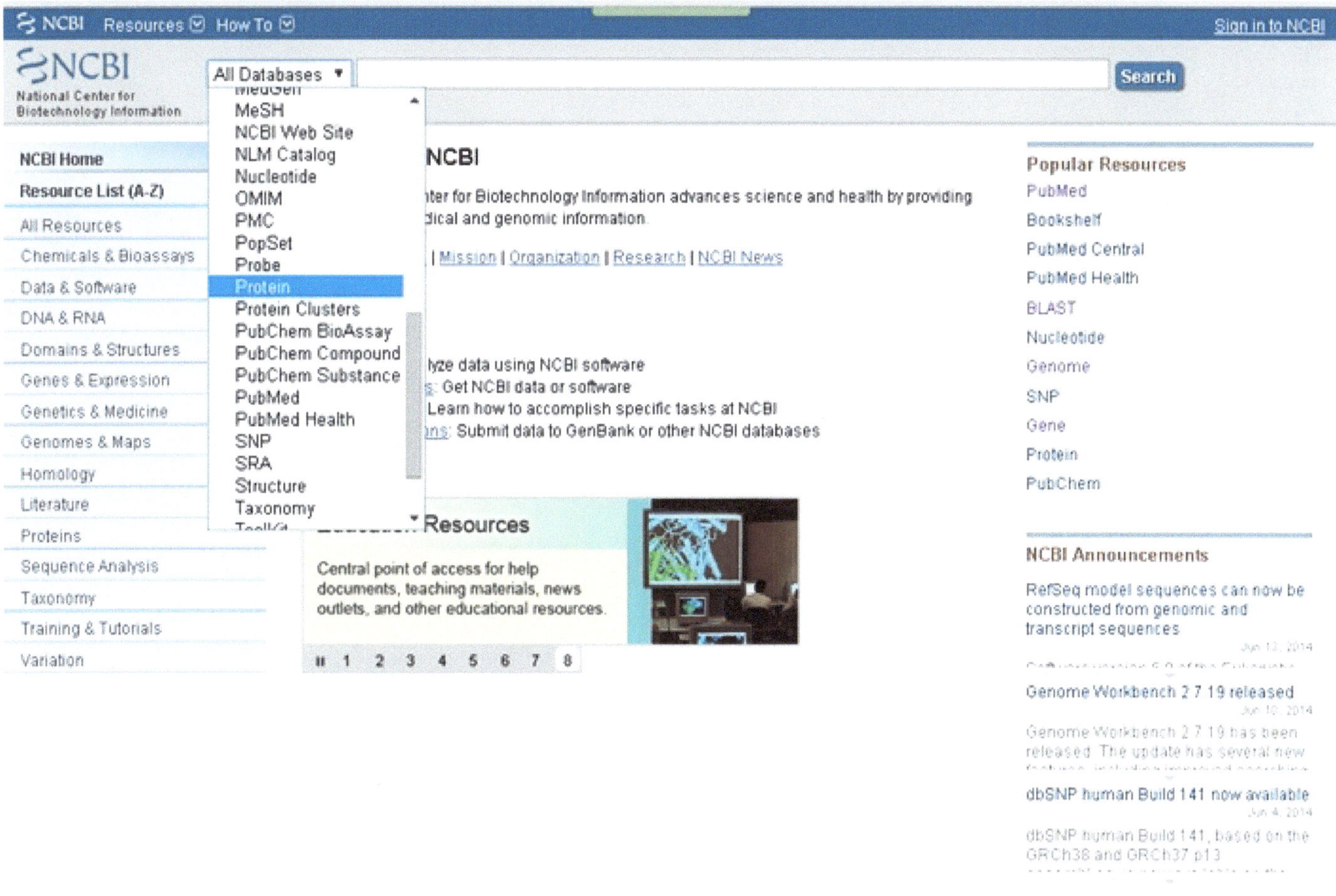

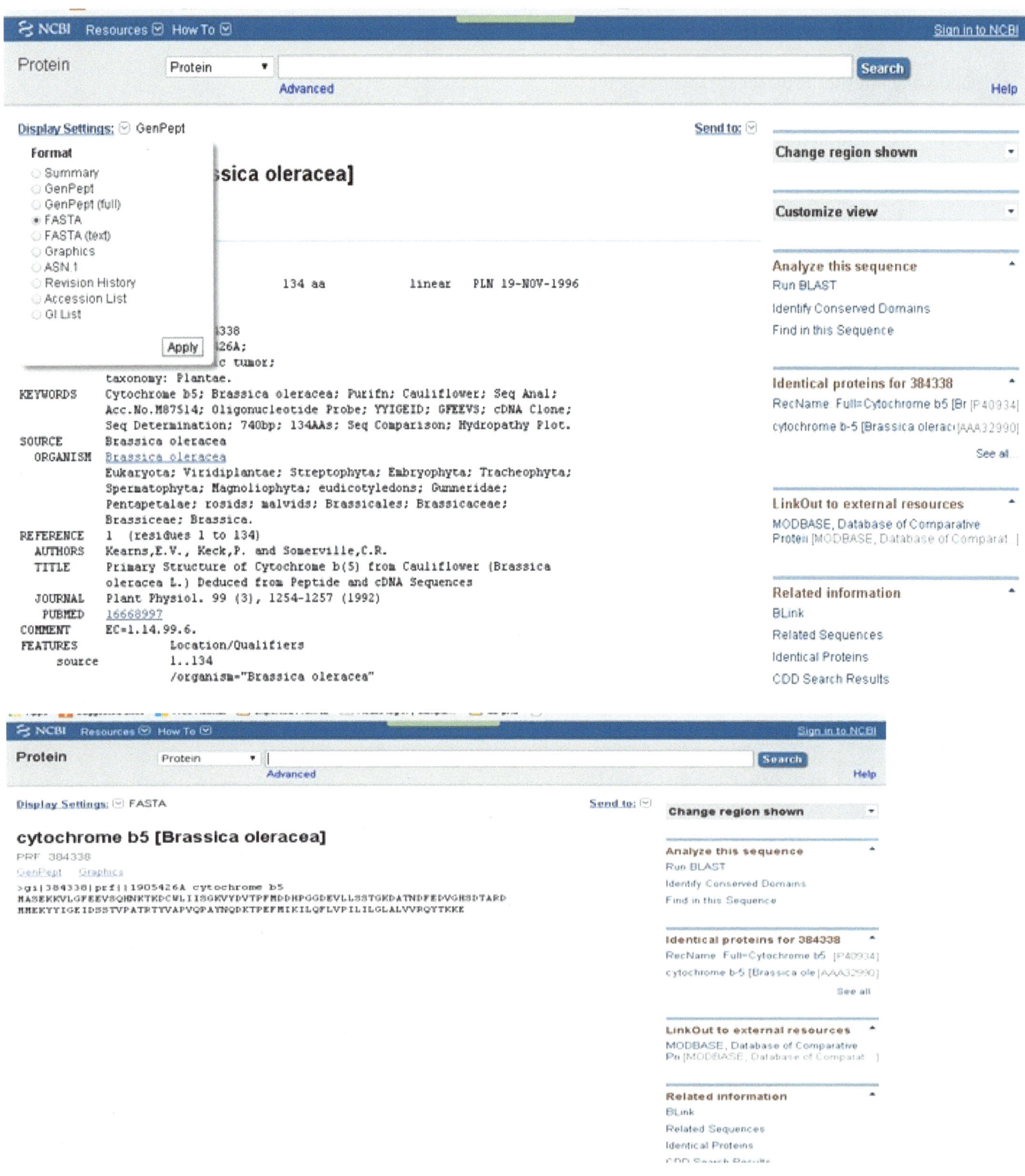
NCBI Resources How To
Sign in to NCBI
Protein
Search
Advanced
Help
Display Settings: GenPept
Send to:
Format
Summary
GenPept
GenPept (full)
FASTA
FASTA (text)
Graphics
ASN.1
Revision History
Accession List
GI List
Apply
sica oleracea]
134 aa linear PLN 19-NOV-1996
taxonomy: Plantae.
KEYWORDS Cytochrome b5; Brassica oleracea; Purifn; Cauliflower; Seq Anal; Acc.No.M87514; Oligonucleotide Probe; YYIGEID; GFEEVS; cDNA Clone; Seq Determination; 740bp; 134AAs; Seq Comparison; Hydropathy Plot.
SOURCE Brassica oleracea
ORGANISM Brassica oleracea
Eukaryota; Viridiplantae; Streptophyta; Embryophyta; Tracheophyta; Spermatophyta; Magnoliophyta; eudicotyledons; Gunneridae; Pentapetalae; rosids; malvids; Brassicales; Brassicaceae; Brassiceae; Brassica.
REFERENCE 1 (residues 1 to 134)
AUTHORS Kearns,E.V., Keck,P. and Somerville,C.R.
TITLE Primary Structure of Cytochrome b(5) from Cauliflower (Brassica oleracea L.) Deduced from Peptide and cDNA Sequences
JOURNAL Plant Physiol. 99 (3), 1254-1257 (1992)
PUBMED 16668997
COMMENT EC=1.14.99.6.
FEATURES Location/Qualifiers
source 1..134
/organism="Brassica oleracea"
Change region shown
Customize view
Analyze this sequence
Run BLAST
Identify Conserved Domains
Find in this Sequence
Identical proteins for 384338
RecName: Full=Cytochrome b5 [Br [P40934]
cytochrome b-5 [Brassica olerac [AAA32990]
See all
LinkOut to external resources
MODBASE, Database of Comparative Protein [MODBASE, Database of Comparat]
Related information
BLink
Related Sequences
Identical Proteins
CDD Search Results
Display Settings: FASTA
cytochrome b5 [Brassica oleracea]
PRF 384338
GenPept Graphics
>gi|384338|prf||1905426A cytochrome b5
MASEKKVLGFEEVSQHNKTKDCWLIISGKVYDVTPFMDDHPGGDEVLLSSTGKDATNDFEDVGHSDTARD
MMEKYYIGEIDSSTVPATRTYVAPVQPAYNQDKTPEFMIKILQFLVPILILGLALVVRQYTKKE

3. RESTRICTION MAPPING

- Open internet browser with link http://www.restrictionmapper.org/
- Paste Nucleotide sequence in the sequence info text box and name your sequence
    - *do not paste sequence in FASTA format , paste only sequence text.
- Select individual enzyme in the Include column.* if you select individual enzyme the result will display
    - For selected enzyme .
- Click Map site in Menu column it will display the result page for selected enzyme for the submitted sequence.
- The result page will display restriction mapping cut position of individual enzyme for respective sequence .

Welcome to RestrictionMapper - on line restriction mapping the easy way.

Maps sites for restriction enzymes, a.k.a. restriction endonucleases, in DNA sequences. Also does virtual digestion.

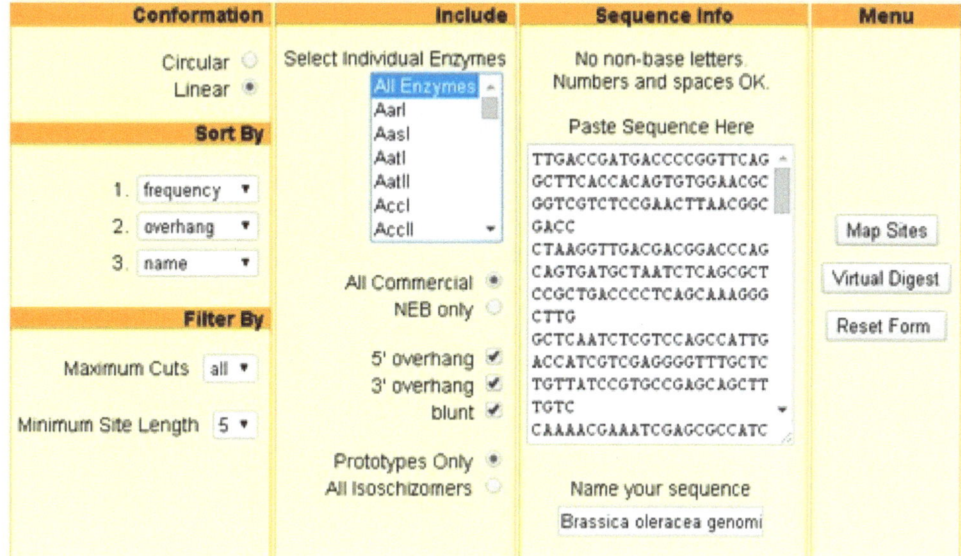

Name: Brassica oleracea genomic

Conformation: linear

Overhang: five_prime, three_prime, blunt

Minimum Site Length: 5 bases

Maximum Number of Cuts: all

Included: all commercial, prototypes only

Noncutters: AatII, AclI, AflII, AflIII, AjuI, AlfI, AlwNI, ApaLI, ArsI, AscI, AsuII, AvrII, BamHI, BarI, BciVI, BfiI, BglII, BplI, BsaAI, BsaXI, BsgI, Bsp1407I, BspHI, BsrI, BsrBI, BsrDI, BstXI, ClaI, CspCI, Eam1105I, Eco57I, EcoNI, EcoRII, EcoRV, FalI, FseI, FspAI, HindIII, HpaI, KpnI, MauBI, MfeI, MluI, NcoI, NdeI, NheI, NotI, NspI, PacI, PasI, PfoI, PmaCI, PmeI, PpiI, PpuMI, PsiI, PI-PspI, PsrI, PstI, PvuII, RsrII, SacI, SacII, SanDI, SapI, ScaI, PI-SceI, SexAI, SfiI, SgrDI, SnaBI, SpeI, SphI, SrfI, Sse8387I, SspI, StuI, SwaI, TatI, TstI, VspI, XbaI, XcmI, XmnI

Name	Sequence	Site Length	Overhang	Frequency	Cut Positions
BalI	TGGCCA	6	blunt	1	441
BsaBI	GATNNNNATC	6	blunt	1	101
BtrI	CACGTC	6	blunt	1	798
NaeI	GCCGGC	6	blunt	1	308
NruI	TCGCGA	6	blunt	1	1054
PshAI	GACNNNNGTC	6	blunt	1	166
SmaI	CCCGGG	6	blunt	1	689
AarI	CACCTGC	7	five_prime	1	656
AbsI	CCTCGAGG	8	five_prime	1	940

4. Similarity Search Using FASTA TOOL FROM EMBL

- Open any Internet Browser
- Go to the link http://www.ncbi.nlm.nih.gov
- On the top of NCBI home page drag down All database and Select Protein.
- In the search text box type key word of query such as Accession number (or) disease name (or) organism name to retrieve the sequence.
- Click on Search button this will display list of available result match to the given keyword.
- Select the check box in the result page which you want to retrieve the sequence.
- Go to display setting and select FASTA format .Because most of the bioinformatics tools accept FASTA file format.
- The result will display in FASTA format.

- Save the file by save as in the top right corner by clicking button send file
    - Or Copy the file text from '>' (greater symbol) up to the End of the sequence text and save in notepad with file name and extension as **.FASTA .**
- Open FASTA tool http://www.ebi.ac.uk/Tools/sss/fasta/
- Step 1-In the FASTA tool webpage select the Nucleotide (or) Protein Database as backend to hit your search
- Step-2 In the FASTA Tool webpage paste the retrieved nucleotide (or)protein sequence in the INPUT SEQUENCE text box (or) Upload your FASTA format file.
- Step 3- set your parameter by selecting different FASTA program in the drop down box.
- Step4- select the check box if you want to receive result by email (or) submit it will display the result in another window

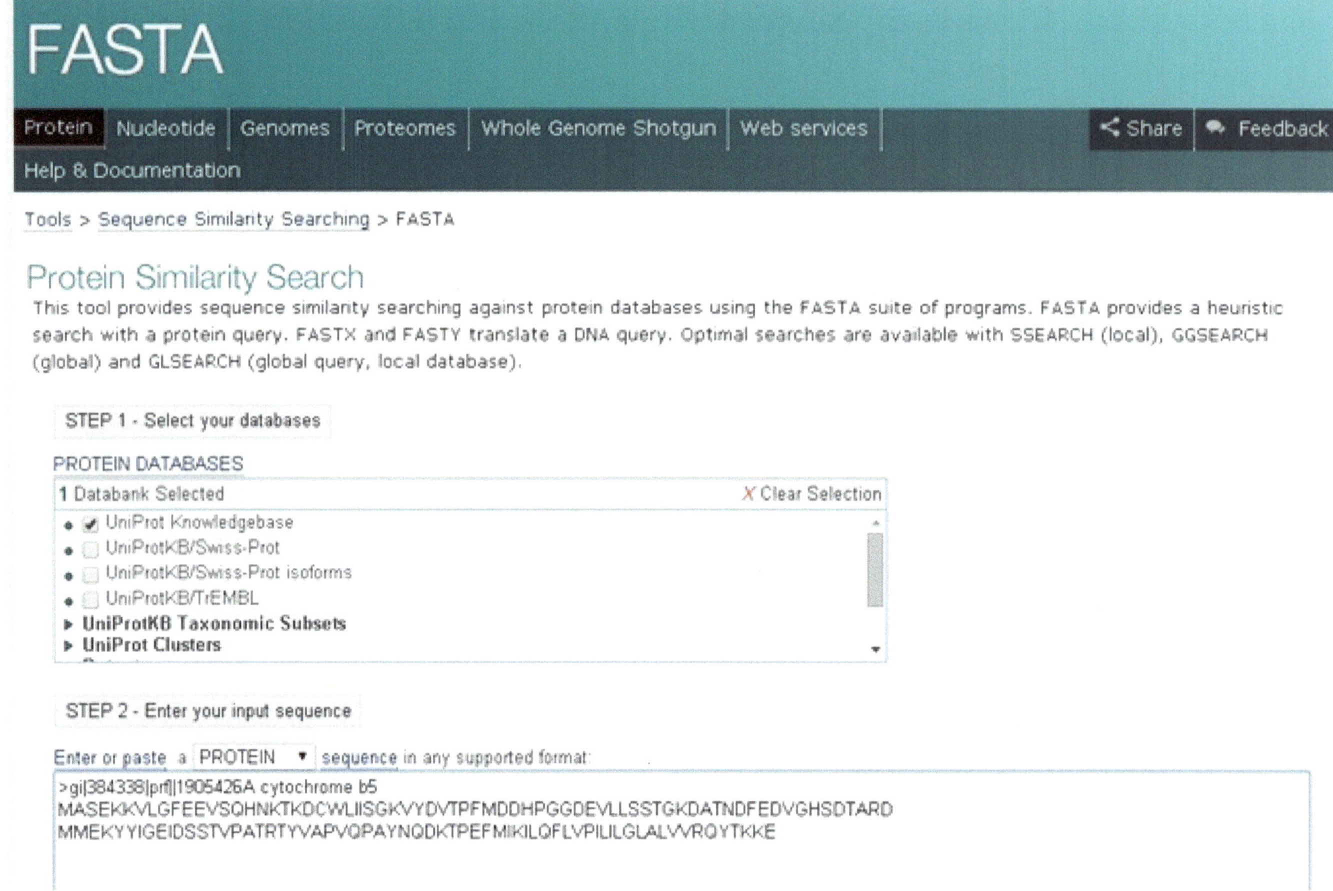
FASTA
Protein
Nucleotide
Genomes
Proteomes
Whole Genome Shotgun
Web services
Share
Feedback
Help & Documentation
Tools > Sequence Similarity Searching > FASTA
Protein Similarity Search
This tool provides sequence similarity searching against protein databases using the FASTA suite of programs. FASTA provides a heuristic search with a protein query. FASTX and FASTY translate a DNA query. Optimal searches are available with SSEARCH (local), GGSEARCH (global) and GLSEARCH (global query, local database).
STEP 1 - Select your databases
PROTEIN DATABASES
1 Databank Selected
X Clear Selection
UniProt Knowledgebase
UniProtKB/Swiss-Prot
UniProtKB/Swiss-Prot isoforms
UniProtKB/TrEMBL
UniProtKB Taxonomic Subsets
UniProt Clusters
STEP 2 - Enter your input sequence
Enter or paste a PROTEIN sequence in any supported format:
>gi|384338|prf||1905426A cytochrome b5
MASEKKVLGFEEVSQHNKTKDCWLIISGKVYDVTPFMDDHPGGDEVLLSSTGKDATNDFEDVGHSDTARD
MMEKYYIGEIDSSTVPATRTYVAPVQPAYNQDKTPEFMIKILQFLVPILILGLALVVRQYTKKE

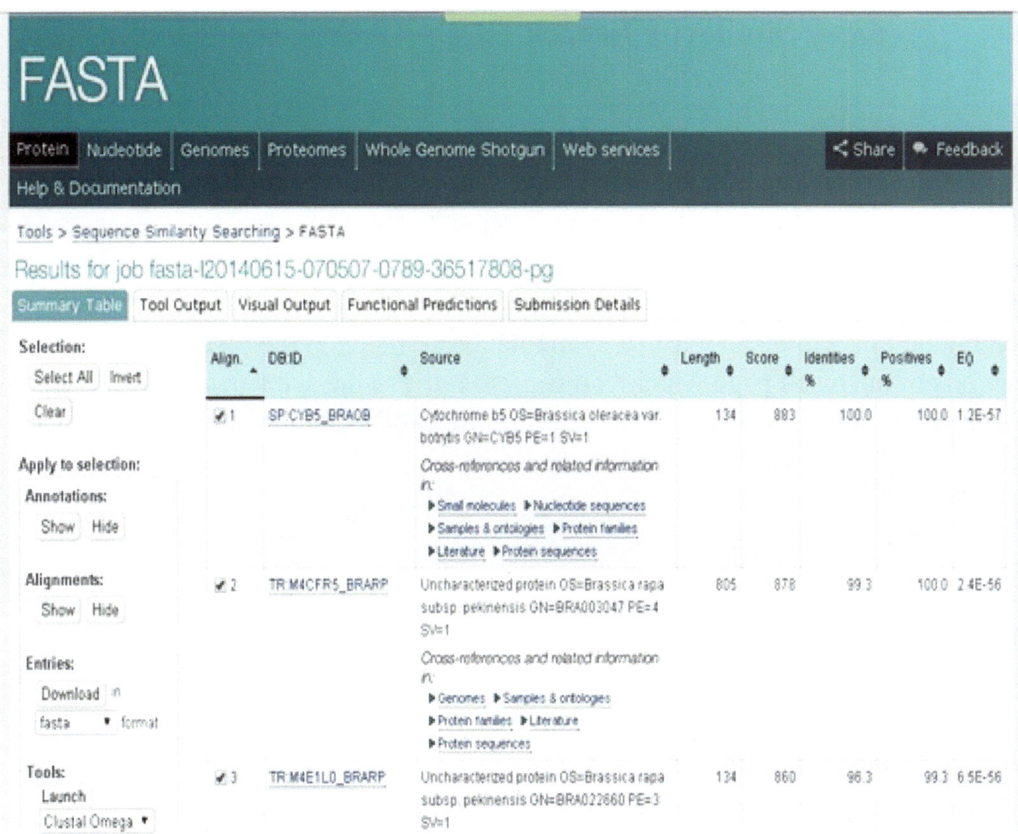
FASTA
Protein
Nucleotide
Genomes
Proteomes
Whole Genome Shotgun
Web services
Share
Feedback
Help & Documentation
Tools > Sequence Similarity Searching > FASTA
Results for job fasta-I20140615-070507-0789-36517808-pg
Summary Table
Tool Output
Visual Output
Functional Predictions
Submission Details
Selection:
Select All
Invert
Clear
Apply to selection:
Annotations:
Show
Hide
Alignments:
Show
Hide
Entries:
Download
fasta
Tools:
Launch
Clustal Omega
Align.
DB:ID
Source
Length
Score
Identities %
Positives %
E()
1
SP:CYB5_BRAOB
Cytochrome b5 OS=Brassica oleracea var. botrytis GN=CYB5 PE=1 SV=1
134
883
100.0
100.0
1.2E-57
Cross-references and related information in:
Small molecules
Nucleotide sequences
Samples & ontologies
Protein families
Literature
Protein sequences
2
TR:M4CFR5_BRARP
Uncharacterized protein OS=Brassica rapa subsp. pekinensis GN=BRA003047 PE=4 SV=1
805
878
99.3
100.0
2.4E-56
Cross-references and related information in:
Genomes
Samples & ontologies
Protein families
Literature
Protein sequences
3
TR:M4E1L0_BRARP
Uncharacterized protein OS=Brassica rapa subsp. pekinensis GN=BRA022660 PE=3 SV=1
134
860
96.3
99.3
6.5E-56

5. Similarity Search Using BLAST TOOL

- Open any Internet Browser
- Go to the link http://www.ncbi.nlm.nih.gov
- On the top of NCBI home page drag down All database and Select Protein.
- In the search text box type key word of query such as Accession number (or) disease name (or) organism name to retrieve the sequence.
- Click on Search button this will display list of available result match to the given keyword.
- Select the check box in the result page which you want to retrieve the sequence.
- Go to display setting and select FASTA format .Because most of the bioinformatics tools accept FASTA file format.
- The result will display in FASTA format

- Save the file by save as in the top right corner by clicking button send file
  - Or Copy the file text from '>' (greater symbol) up to the End of the sequence text and save in notepad with file name and extension as **.FASTA .**
- Open the Blast home page from NCBI Home page http://www.ncbi.nlm.nih.gov Blast link in the right side.
- Blast Home page link http://blast.ncbi.nlm.nih.gov/Blast.cgi
- Select Protein BLAST if submitting protein Sequence.
- Paste your FASTA sequence or Accession No in the text box (or) Upload FASTA format file .
- Click the **BLAST** button to get the output.

BLAST

The given Organism Brassica oleracea , Accession : 1905426A
gi|384338|BLAST of Protein sequence for BLAST.

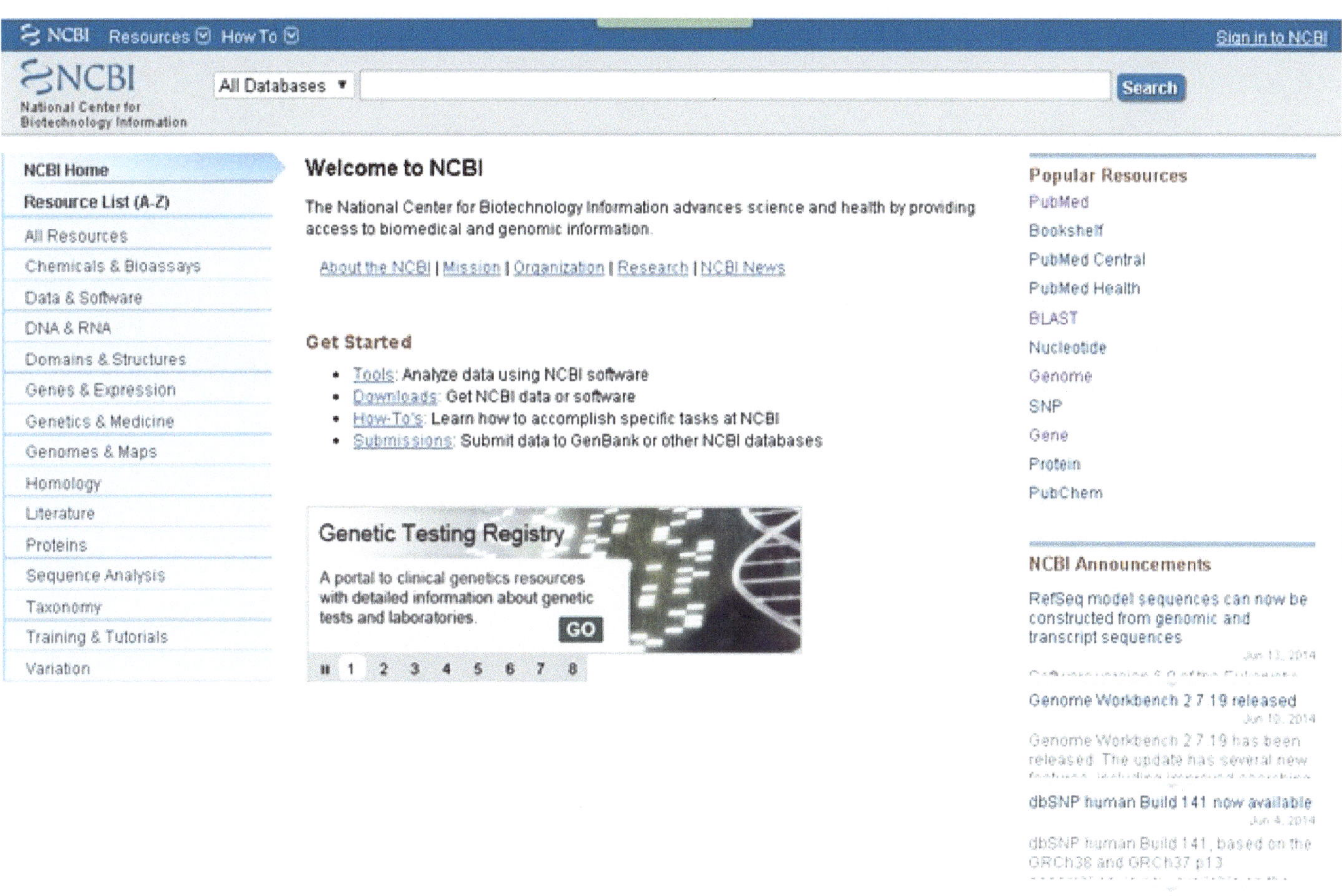

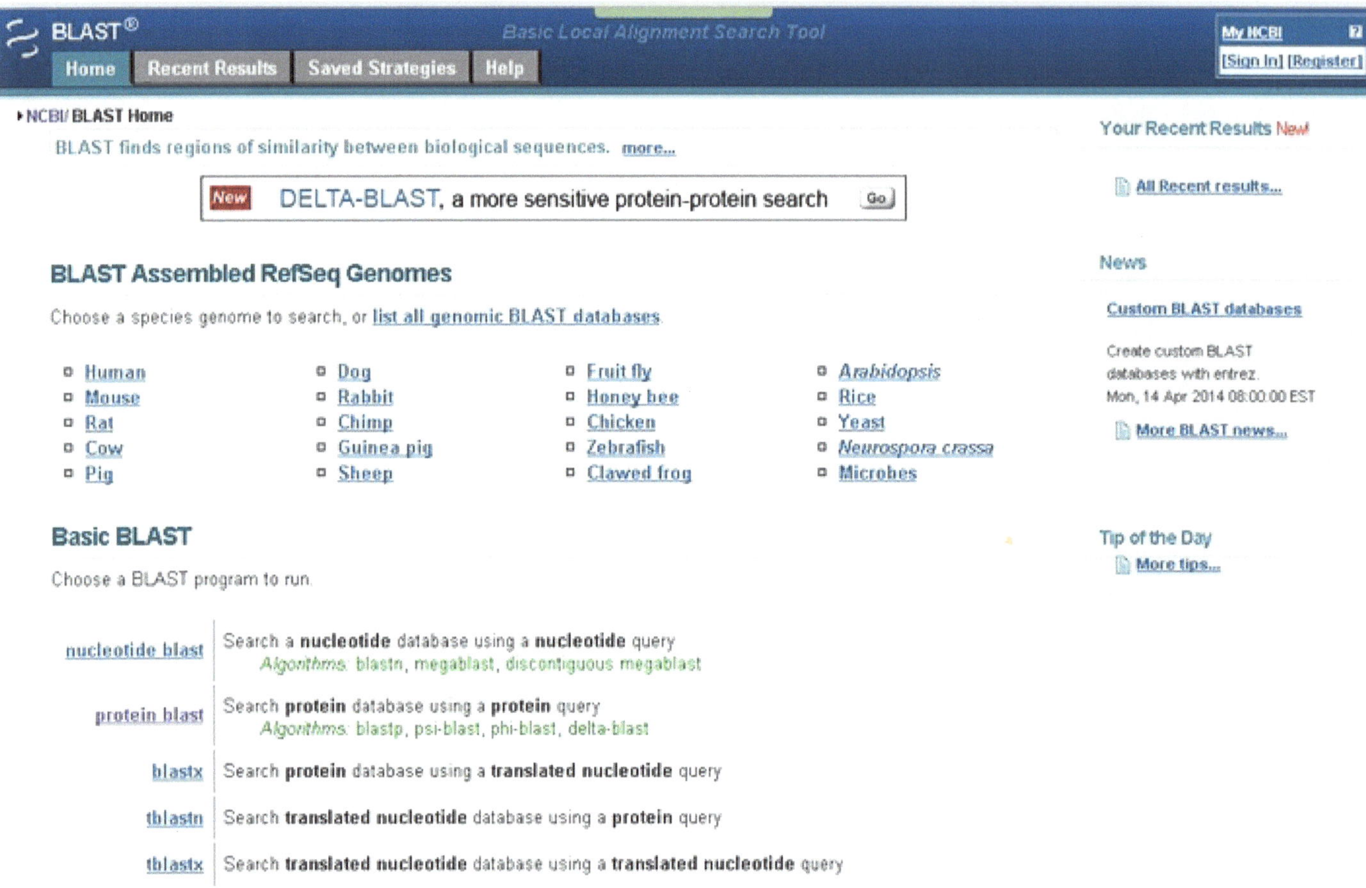
BLAST®
Basic Local Alignment Search Tool
Home
Recent Results
Saved Strategies
Help
My NCBI
[Sign In] [Register]
NCBI/ BLAST Home
BLAST finds regions of similarity between biological sequences. more...
New DELTA-BLAST, a more sensitive protein-protein search Go
BLAST Assembled RefSeq Genomes
Choose a species genome to search, or list all genomic BLAST databases.
Human
Mouse
Rat
Cow
Pig
Dog
Rabbit
Chimp
Guinea pig
Sheep
Fruit fly
Honey bee
Chicken
Zebrafish
Clawed frog
Arabidopsis
Rice
Yeast
Neurospora crassa
Microbes
Basic BLAST
Choose a BLAST program to run.
nucleotide blast
Search a nucleotide database using a nucleotide query
Algorithms: blastn, megablast, discontiguous megablast
protein blast
Search protein database using a protein query
Algorithms: blastp, psi-blast, phi-blast, delta-blast
blastx
Search protein database using a translated nucleotide query
tblastn
Search translated nucleotide database using a protein query
tblastx
Search translated nucleotide database using a translated nucleotide query
Your Recent Results New!
All Recent results...
News
Custom BLAST databases
Create custom BLAST databases with entrez.
Mon, 14 Apr 2014 08:00:00 EST
More BLAST news...
Tip of the Day
More tips...

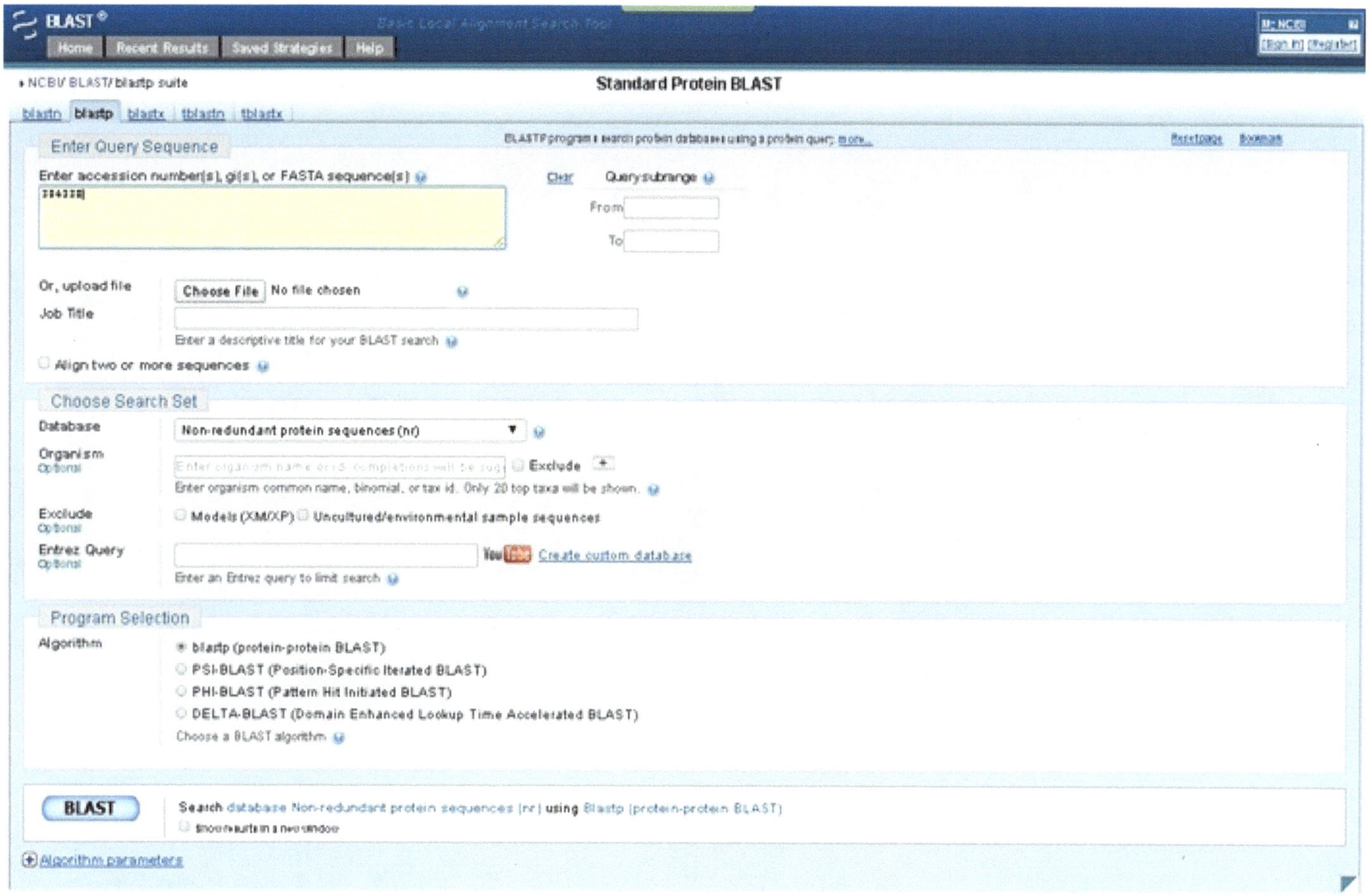

BLAST® Basic Local Alignment Search Tool

Home | Recent Results | Saved Strategies | Help

NCBI/ BLAST/ blastp suite

Standard Protein BLAST

blastn | blastp | blastx | tblastn | tblastx

Enter Query Sequence

BLASTP programs search protein databases using a protein query. more...

Reset page | Bookmark

Enter accession number(s), gi(s), or FASTA sequence(s) Clear

384338

Query subrange From To

Or, upload file Choose File No file chosen

Job Title

Enter a descriptive title for your BLAST search

Align two or more sequences

Choose Search Set

Database Non-redundant protein sequences (nr)

Organism Optional Exclude

Enter organism common name, binomial, or tax id. Only 20 top taxa will be shown.

Exclude Optional Models (XM/XP) Uncultured/environmental sample sequences

Entrez Query Optional Create custom database

Enter an Entrez query to limit search

Program Selection

Algorithm

- blastp (protein-protein BLAST)
- PSI-BLAST (Position-Specific Iterated BLAST)
- PHI-BLAST (Pattern Hit Initiated BLAST)
- DELTA-BLAST (Domain Enhanced Lookup Time Accelerated BLAST)

Choose a BLAST algorithm

BLAST Search database Non-redundant protein sequences (nr) using Blastp (protein-protein BLAST)

Show results in a new window

Algorithm parameters

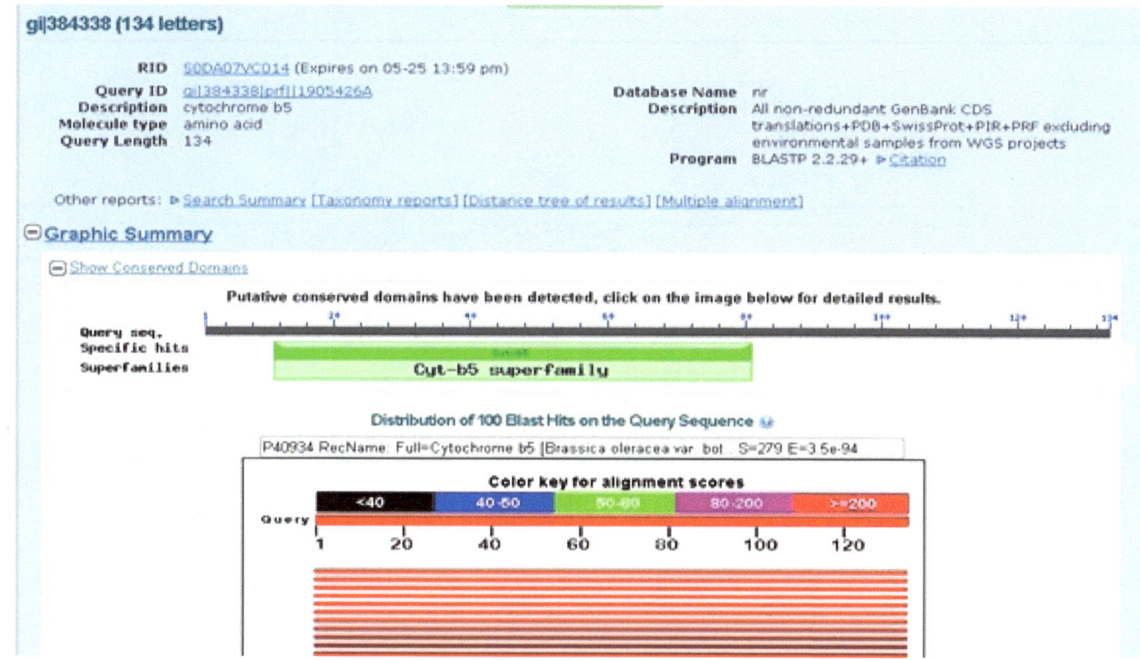

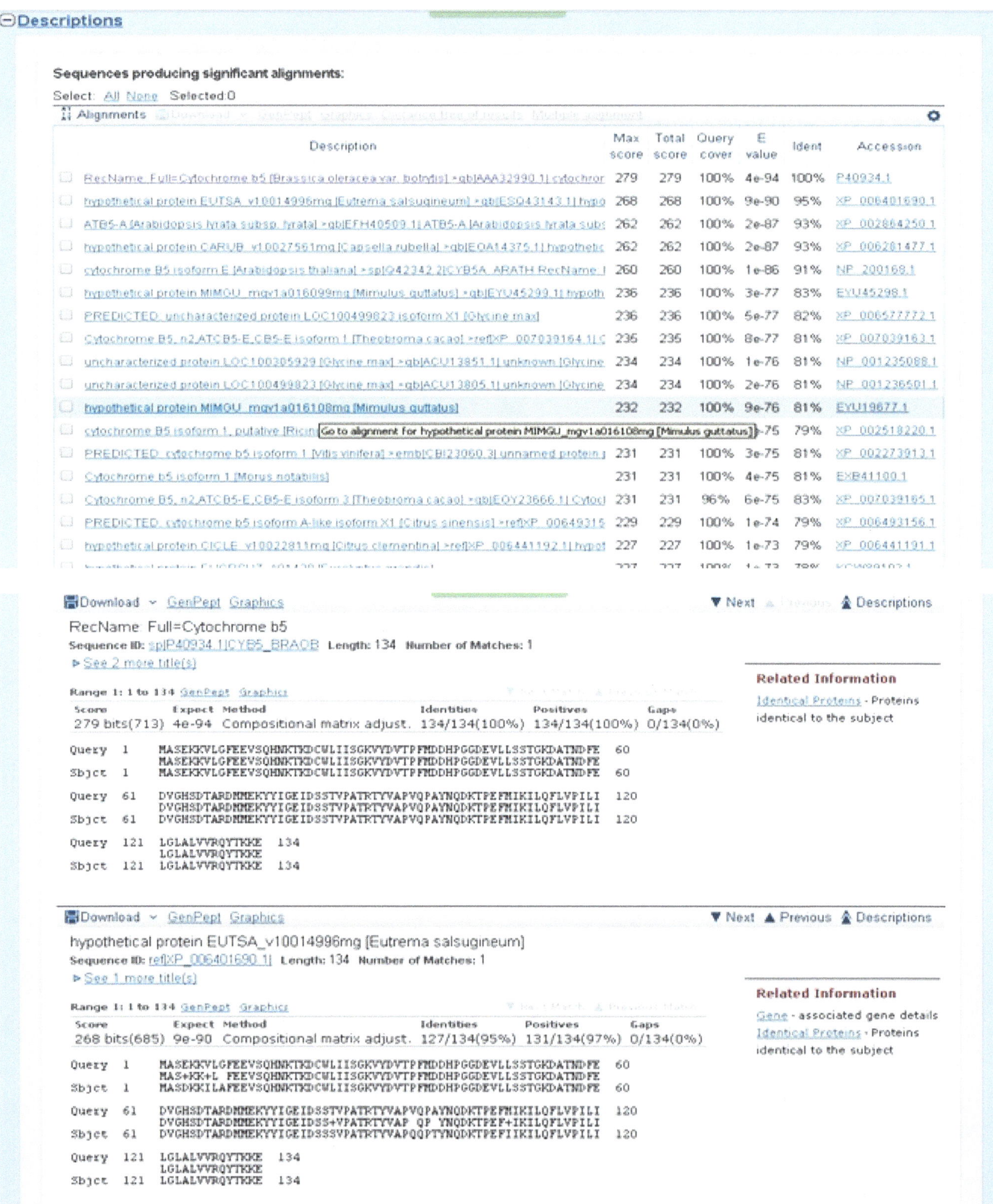

⊖Descriptions

Sequences producing significant alignments:

Select: All None Selected:0

Alignments

	Description	Max score	Total score	Query cover	E value	Ident	Accession
☐	RecName: Full=Cytochrome b5 [Brassica oleracea var. botrytis] >gb|AAA32990.1| cytochror	279	279	100%	4e-94	100%	P40934.1
☐	hypothetical protein EUTSA_v10014996mg [Eutrema salsugineum] >gb|ESQ43143.1| hypo	268	268	100%	9e-90	95%	XP_006401690.1
☐	ATB5-A [Arabidopsis lyrata subsp. lyrata] >gb|EFH40509.1| ATB5-A [Arabidopsis lyrata subs	262	262	100%	2e-87	93%	XP_002864250.1
☐	hypothetical protein CARUB_v10027561mg [Capsella rubella] >gb|EOA14375.1| hypothetic	262	262	100%	2e-87	93%	XP_006281477.1
☐	cytochrome B5 isoform E [Arabidopsis thaliana] >sp|Q42342.2|CYB5A_ARATH RecName: I	260	260	100%	1e-86	91%	NP_200168.1
☐	hypothetical protein MIMGU_mgv1a016099mg [Mimulus guttatus] >gb|EYU45299.1| hypoth	236	236	100%	3e-77	83%	EYU45298.1
☐	PREDICTED: uncharacterized protein LOC100499823 isoform X1 [Glycine max]	236	236	100%	5e-77	82%	XP_006577772.1
☐	Cytochrome B5, n2,ATCB5-E,CB5-E isoform 1 [Theobroma cacao] >ref|XP_007039164.1| C	235	235	100%	8e-77	81%	XP_007039163.1
☐	uncharacterized protein LOC100305929 [Glycine max] >gb|ACU13851.1| unknown [Glycine	234	234	100%	1e-76	81%	NP_001235088.1
☐	uncharacterized protein LOC100499823 [Glycine max] >gb|ACU13805.1| unknown [Glycine	234	234	100%	2e-76	81%	NP_001236501.1
☐	hypothetical protein MIMGU_mgv1a016108mg [Mimulus guttatus]	232	232	100%	9e-76	81%	EYU19677.1
☐	cytochrome B5 isoform 1, putative [Ricin [Go to alignment for hypothetical protein MIMGU_mgv1a016108mg [Mimulus guttatus]]	[illegible]	[illegible]	[illegible]	[illegible]-75	79%	XP_002518220.1
☐	PREDICTED: cytochrome b5 isoform 1 [Vitis vinifera] >emb|CBI23060.3| unnamed protein ;	231	231	100%	3e-75	81%	XP_002273913.1
☐	Cytochrome b5 isoform 1 [Morus notabilis]	231	231	100%	4e-75	81%	EXB41100.1
☐	Cytochrome B5, n2,ATCB5-E,CB5-E isoform 3 [Theobroma cacao] >gb|EOY23666.1| Cytoc	231	231	96%	6e-75	83%	XP_007039165.1
☐	PREDICTED: cytochrome b5 isoform A-like isoform X1 [Citrus sinensis] >ref|XP_00649315	229	229	100%	1e-74	79%	XP_006493156.1
☐	hypothetical protein CICLE_v10022811mg [Citrus clementina] >ref|XP_006441192.1| hypot	227	227	100%	1e-73	79%	XP_006441191.1

Download GenPept Graphics ▼ Next ▲ Previous Descriptions

RecName: Full=Cytochrome b5

Sequence ID: sp|P40934.1|CYB5_BRAOB **Length:** 134 **Number of Matches:** 1

▶ See 2 more title(s)

Range 1: 1 to 134 GenPept Graphics

Score	Expect	Method	Identities	Positives	Gaps
279 bits(713)	4e-94	Compositional matrix adjust.	134/134(100%)	134/134(100%)	0/134(0%)

```
Query  1    MASEKKVLGFEEVSQHNKTKDCWLIISGKVYDVTPFMDDHPGGDEVLLSSTGKDATNDFE  60
            MASEKKVLGFEEVSQHNKTKDCWLIISGKVYDVTPFMDDHPGGDEVLLSSTGKDATNDFE
Sbjct  1    MASEKKVLGFEEVSQHNKTKDCWLIISGKVYDVTPFMDDHPGGDEVLLSSTGKDATNDFE  60

Query  61   DVGHSDTARDMMEKYYIGEIDSSTVPATRTYVAPVQPAYNQDKTPEFMIKILQFLVPILI  120
            DVGHSDTARDMMEKYYIGEIDSSTVPATRTYVAPVQPAYNQDKTPEFMIKILQFLVPILI
Sbjct  61   DVGHSDTARDMMEKYYIGEIDSSTVPATRTYVAPVQPAYNQDKTPEFMIKILQFLVPILI  120

Query  121  LGLALVVRQYTKKE  134
            LGLALVVRQYTKKE
Sbjct  121  LGLALVVRQYTKKE  134
```

Related Information

Identical Proteins - Proteins identical to the subject

Download GenPept Graphics ▼ Next ▲ Previous Descriptions

hypothetical protein EUTSA_v10014996mg [Eutrema salsugineum]

Sequence ID: ref|XP_006401690.1| **Length:** 134 **Number of Matches:** 1

▶ See 1 more title(s)

Range 1: 1 to 134 GenPept Graphics

Score	Expect	Method	Identities	Positives	Gaps
268 bits(685)	9e-90	Compositional matrix adjust.	127/134(95%)	131/134(97%)	0/134(0%)

```
Query  1    MASEKKVLGFEEVSQHNKTKDCWLIISGKVYDVTPFMDDHPGGDEVLLSSTGKDATNDFE  60
            MAS+KK+L FEEVSQHNKTKDCWLIISGKVYDVTPFMDDHPGGDEVLLSSTGKDATNDFE
Sbjct  1    MASDKKILAFEEVSQHNKTKDCWLIISGKVYDVTPFMDDHPGGDEVLLSSTGKDATNDFE  60

Query  61   DVGHSDTARDMMEKYYIGEIDSSTVPATRTYVAPVQPAYNQDKTPEFMIKILQFLVPILI  120
            DVGHSDTARDMMEKYYIGEIDSS+VPATRTYVAP QP YNQDKTPEF+IKILQFLVPILI
Sbjct  61   DVGHSDTARDMMEKYYIGEIDSSSVPATRTYVAPQQPTYNQDKTPEFIIKILQFLVPILI  120

Query  121  LGLALVVRQYTKKE  134
            LGLALVVRQYTKKE
Sbjct  121  LGLALVVRQYTKKE  134
```

Related Information

Gene - associated gene details

Identical Proteins - Proteins identical to the subject

6. PAIRWISE SEQUENCE ALIGNEMNT –EMBOSS SUITE

- Go to http://www.ebi.ac.uk/Tools/emboss/
- Emboss suite is a tool to compare any two sequence with the database.
- If you want to align the whole length of both the sequence use EMBOS NEEDLE.
- If you want to find the best region of similarity of both the sequence use EMBOS WATER -MAN
- Go to http://www.ebi.ac.uk/Tools/emboss/
- If you are submitting protein sequence select Protein under NEEDLE (or)WATER MAN option from this EMBOSS SUITE page
- If you are submitting nucleotide sequence select nucleotide under NEEDLE (or)WATER MAN option from this EMBOSS SUITE page.
- Download any two FASTA format sequence of your interest from NCBI database.
- Paste the two sequence in the Align home page i.e http://www.ebi.ac.uk/Tools/psa/emboss_needle/
    - In two separate window.
- Select the type of alignment you want either local or global
- Choose the molecule either nucleic acid or protein.
- Choose the matrix which you wish to use
- Click on run to get the result.

Pairwise Alignment

The given two Organism
1.Brassica oleracea , Accession : 1905426A gi|384338|
2. Bacillus Subtilus , PDB : 3F3B gi |215261413|

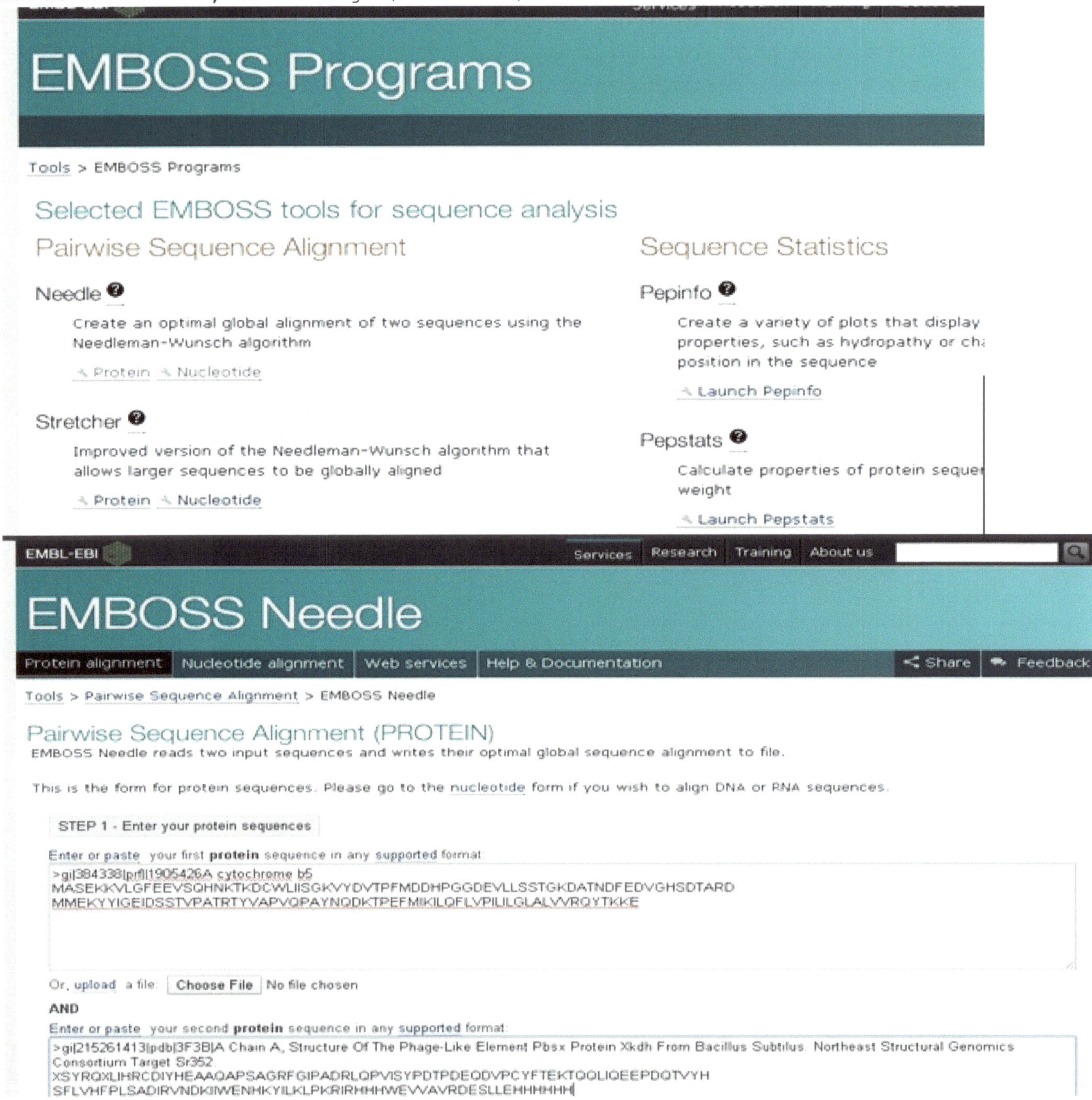

Results for job emboss_needle-I20140524-070951-0246-27710773-es

Alignment | Submission Details

View Alignment File

```
########################################
# Program: needle
# Rundate: Sat 24 May 2014 07:09:52
# Commandline: needle
#    -auto
#    -stdout
#    -asequence emboss_needle-I20140524-070951-0246-27710773-es.asequence
#    -bsequence emboss_needle-I20140524-070951-0246-27710773-es.bsequence
#    -datafile EBLOSUM62
#    -gapopen 10.0
#    -gapextend 0.5
#    -endopen 10.0
#    -endextend 0.5
#    -aformat3 pair
#    -sprotein1
#    -sprotein2
# Align_format: pair
# Report_file: stdout
########################################

#=======================================
#
# Aligned_sequences: 2
# 1: 1905426A
# 2: 3F3BA
# Matrix: EBLOSUM62
# Gap_penalty: 10.0
# Extend_penalty: 0.5
#
# Length: 223
# Identity:       5/223 ( 2.2%)
# Similarity:    15/223 ( 6.7%)
# Gaps:         186/223 (83.4%)
# Score: 8.5
#
#
#=======================================
```

```
# Gaps:         186/223 (83.4%)
# Score: 8.5
#
#
#=======================================

1905426A           1 --------------------------------------------------      0

3F3BA              1 XSYRQXLIHRCDIYHEAAQAPSAGRFGIPADRLQPVISYPDTPDEQDVPC     50

1905426A           1 ---------------------------------------MASEKKVLGFE     11
                                                            ..:.|.:|...
3F3BA             51 YFTEKTQQLIQEEPDQTVYHSFLVHFPLSADIRVNDKIIWENHKYILKLP    100

1905426A          12 EVSQHNKTKDCWLIISGKVYDVTPFMDDHPGGDEVLLSSTGKDATNDFED     61
                     :..:|:.    |.:::.:...:...,...|.
3F3BA            101 KRIRMMM----WEVVAVRDESLLEMMMMMM--------------------    126

1905426A          62 VGHSDTARDMMEKYYIGEIDSSTVPATRTYVAPVQPAYNQDKTPEFMIKI    111

3F3BA            127 --------------------------------------------------    126

1905426A         112 LQFLVPILILGLALVVRQYTKKE    134

3F3BA            127 -----------------------    126


#---------------------------------------
#---------------------------------------
```

7. MULTIPLE SEQUENCE ALIGNMENT – CLUSTAL OMEGA

- Open any Internet Browser
- Go to the link http://www.ncbi.nlm.nih.gov
- On the top of NCBI home page drag down All database and Select Protein.
- In the search text box type key word of query such as Accession number (or) disease name (or) organism name to retrieve the sequence.
- Click on Search button this will display list of available result match to the given keyword.
- Select the check box in the result page which you want to retrieve the sequence.
- Go to display setting and select FASTA format .Because most of the bioinformatics tools accept FASTA file format.
- The result will display in FASTA format

- Save the file by save as in the top right corner by clicking button send file
  - Or Copy the file text from '>' (greater symbol) up to the End of the sequence text and save in notepad with file name and extension as **.FASTA .**
- Open the Blast home page from NCBI Home page http://www.ncbi.nlm.nih.gov Blast link in the right side.
- Blast Home page link http://blast.ncbi.nlm.nih.gov/Blast.cgi
- Select Protein BLAST if submitting protein Sequence.
- Paste your FASTA sequence or Accession No in the text box (or) Upload FASTA format file .
- Click the **BLAST** button to get the output.
- Select any six sequence from the result of the BLAST web page and download
- Open CLUSTAL OMEGA Tool for (Multiple Sequence Alignment) http://www.ebi.ac.uk/Tools/msa/clustalo/
- Paste your set of all sequence in give text box minimum number should be 5 set of FASTA format Sequence (or) upload Downloaded file form BLAST output and click the submit button.
- The result of the CLUSTAL OMEGA is displayed.

Result : The given 6 set of gi 's such as

gi|604346994|gb|EYU45298.1| 0.12966
gi|565436025|ref|XP_006281477.1| 0.01959
gi|297792731|ref|XP_002864250.1| 0.00622
gi|15238776|ref|NP_200168.1| 0.01617
gi|729252|sp|P40934.1|CYB5_BRAOB 0.03078
gi|567179748|ref|XP_006401690.1| 0.02146

is submitted in CLUSTAL OMEGA Tool for MSA Alignment and the Result for submitted sequence is found

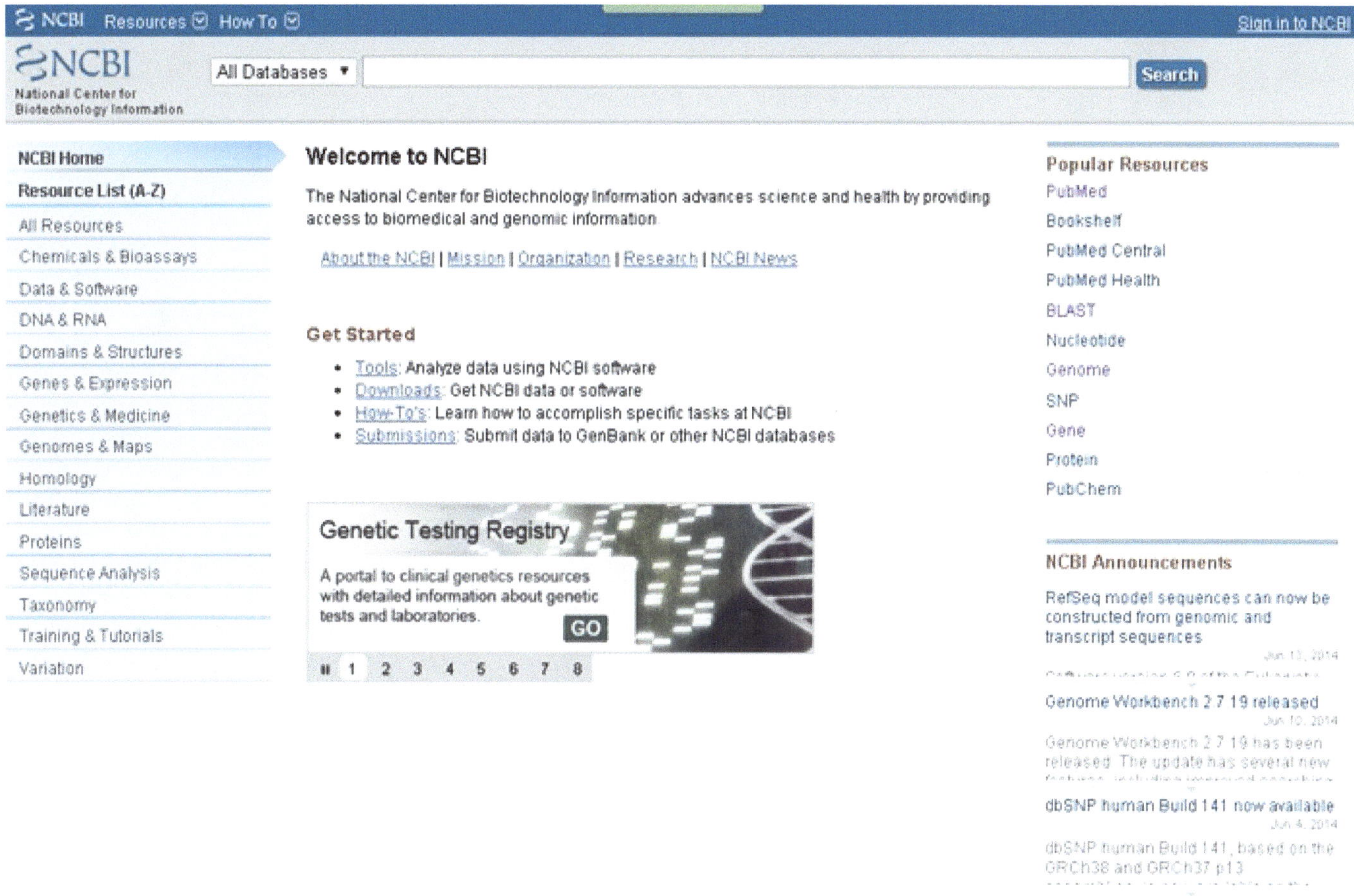

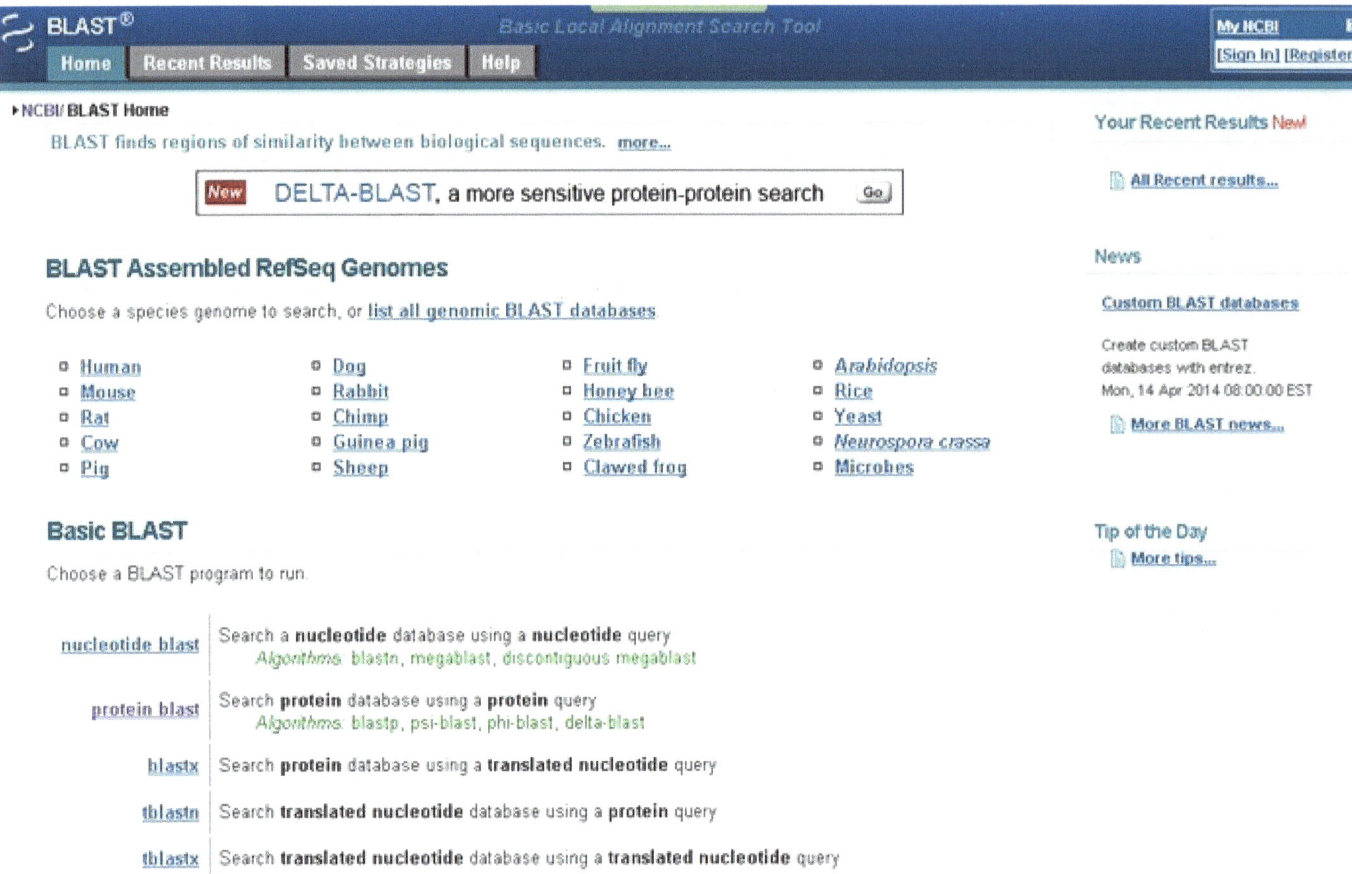
BLAST®
Basic Local Alignment Search Tool
Home
Recent Results
Saved Strategies
Help
My NCBI
[Sign In] [Register]
NCBI/ BLAST Home
BLAST finds regions of similarity between biological sequences. more...
New DELTA-BLAST, a more sensitive protein-protein search Go
BLAST Assembled RefSeq Genomes
Choose a species genome to search, or list all genomic BLAST databases.
Human
Mouse
Rat
Cow
Pig
Dog
Rabbit
Chimp
Guinea pig
Sheep
Fruit fly
Honey bee
Chicken
Zebrafish
Clawed frog
Arabidopsis
Rice
Yeast
Neurospora crassa
Microbes
Basic BLAST
Choose a BLAST program to run.
nucleotide blast
Search a nucleotide database using a nucleotide query
Algorithms: blastn, megablast, discontiguous megablast
protein blast
Search protein database using a protein query
Algorithms: blastp, psi-blast, phi-blast, delta-blast
blastx
Search protein database using a translated nucleotide query
tblastn
Search translated nucleotide database using a protein query
tblastx
Search translated nucleotide database using a translated nucleotide query
Your Recent Results New!
All Recent results...
News
Custom BLAST databases
Create custom BLAST databases with entrez.
Mon, 14 Apr 2014 08:00:00 EST
More BLAST news...
Tip of the Day
More tips...

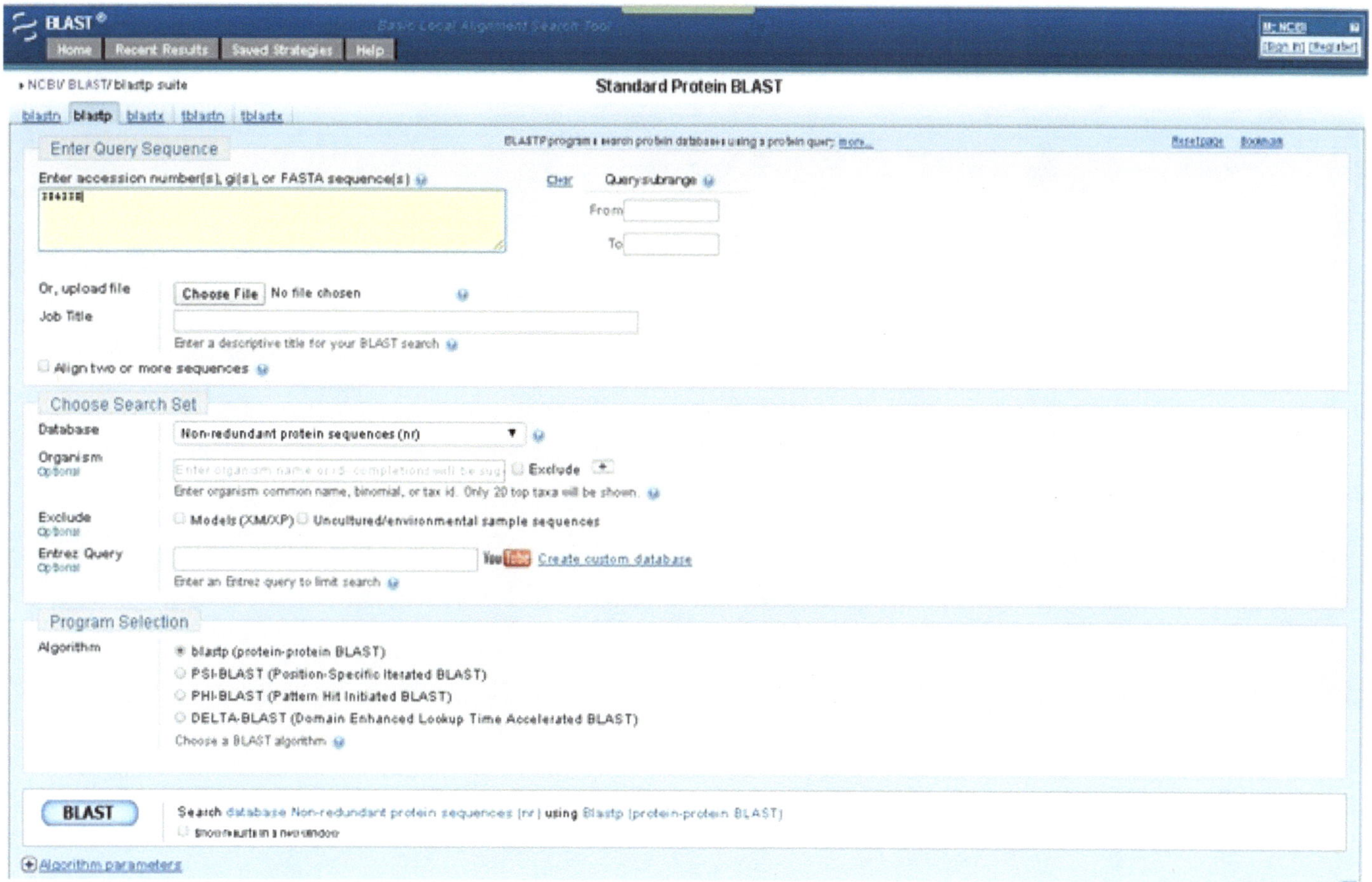

BLAST®
Basic Local Alignment Search Tool
Home | Recent Results | Saved Strategies | Help

NCBI/ BLAST/ blastp suite

Standard Protein BLAST

blastn | blastp | blastx | tblastn | tblastx

Enter Query Sequence

BLASTP programs search protein databases using a protein query. more...

Reset page | Bookmark

Enter accession number(s), gi(s), or FASTA sequence(s)

Clear

Query subrange

From

To

Or, upload file — Choose File — No file chosen

Job Title

Enter a descriptive title for your BLAST search

Align two or more sequences

Choose Search Set

Database — Non-redundant protein sequences (nr)

Organism (Optional) — Exclude

Enter organism common name, binomial, or tax id. Only 20 top taxa will be shown.

Exclude (Optional) — Models (XM/XP) — Uncultured/environmental sample sequences

Entrez Query (Optional) — Create custom database

Enter an Entrez query to limit search

Program Selection

Algorithm

- blastp (protein-protein BLAST)
- PSI-BLAST (Position-Specific Iterated BLAST)
- PHI-BLAST (Pattern Hit Initiated BLAST)
- DELTA-BLAST (Domain Enhanced Lookup Time Accelerated BLAST)

Choose a BLAST algorithm

BLAST — Search database Non-redundant protein sequences (nr) using Blastp (protein-protein BLAST)

Show results in a new window

Algorithm parameters

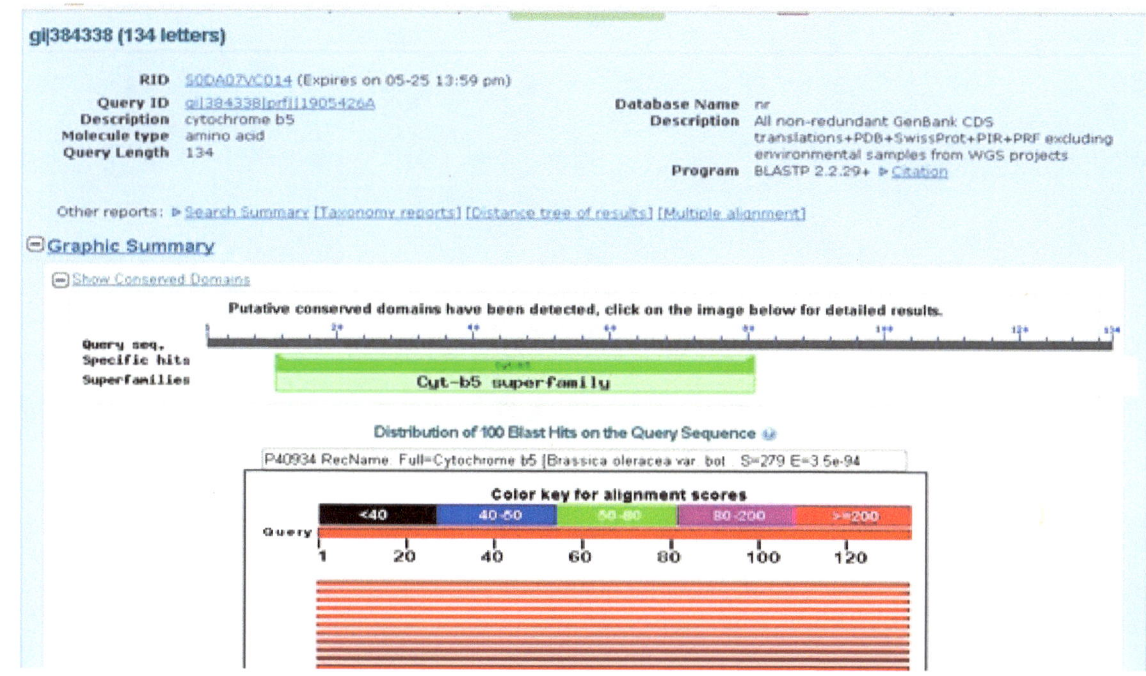

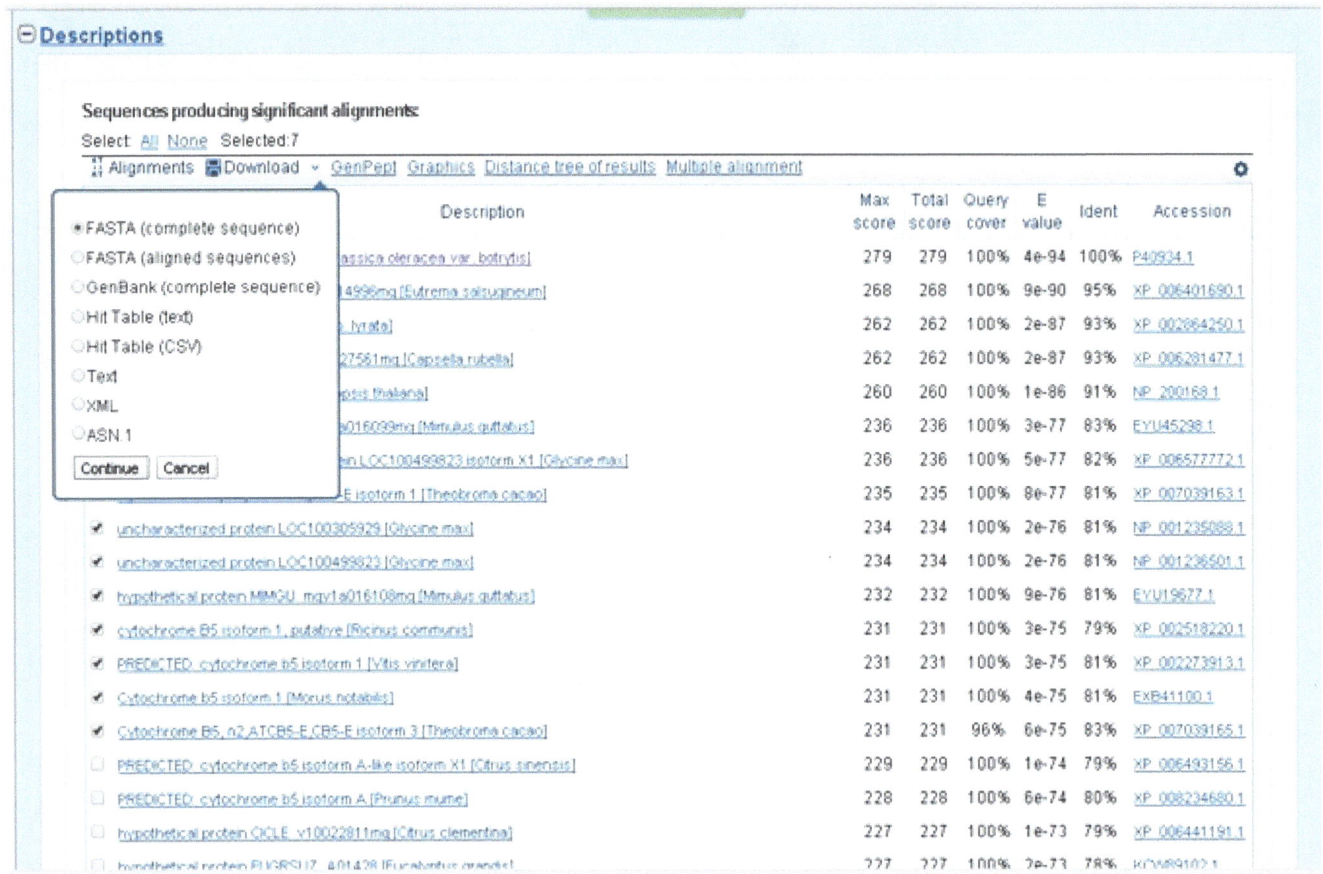

Descriptions

Sequences producing significant alignments:

Select: All None Selected:7

Alignments Download GenPept Graphics Distance tree of results Multiple alignment

FASTA (complete sequence)
FASTA (aligned sequences)
GenBank (complete sequence)
Hit Table (text)
Hit Table (CSV)
Text
XML
ASN.1
Continue Cancel

Description	Max score	Total score	Query cover	E value	Ident	Accession
assica oleracea var. botrytis]	279	279	100%	4e-94	100%	P40934.1
14996mg [Eutrema salsugineum]	268	268	100%	9e-90	95%	XP_006401690.1
lyrata]	262	262	100%	2e-87	93%	XP_002864250.1
27561mg [Capsella rubella]	262	262	100%	2e-87	93%	XP_006281477.1
psis thaliana]	260	260	100%	1e-86	91%	NP_200168.1
a016099mg [Mimulus guttatus]	236	236	100%	3e-77	83%	EYU45298.1
in LOC100499823 isoform X1 [Glycine max]	236	236	100%	5e-77	82%	XP_006577772.1
E isoform 1 [Theobroma cacao]	235	235	100%	8e-77	81%	XP_007039163.1
uncharacterized protein LOC100305929 [Glycine max]	234	234	100%	2e-76	81%	NP_001235088.1
uncharacterized protein LOC100499823 [Glycine max]	234	234	100%	2e-76	81%	NP_001236501.1
hypothetical protein MIMGU_mgv1a016108mg [Mimulus guttatus]	232	232	100%	9e-76	81%	EYU19677.1
cytochrome B5 isoform 1, putative [Ricinus communis]	231	231	100%	3e-75	79%	XP_002518220.1
PREDICTED: cytochrome b5 isoform 1 [Vitis vinifera]	231	231	100%	3e-75	81%	XP_002273913.1
Cytochrome b5 isoform 1 [Morus notabilis]	231	231	100%	4e-75	81%	EXB41100.1
Cytochrome B5, n2,ATCB5-E,CB5-E isoform 3 [Theobroma cacao]	231	231	96%	6e-75	83%	XP_007039165.1
PREDICTED: cytochrome b5 isoform A-like isoform X1 [Citrus sinensis]	229	229	100%	1e-74	79%	XP_006493156.1
PREDICTED: cytochrome b5 isoform A [Prunus mume]	228	228	100%	6e-74	80%	XP_008234680.1
hypothetical protein CICLE_v10022811mg [Citrus clementina]	227	227	100%	1e-73	79%	XP_006441191.1
hypothetical protein EUGRSUZ_A01428 [Eucalyptus grandis]	227	227	100%	2e-73	78%	KCW89102.1

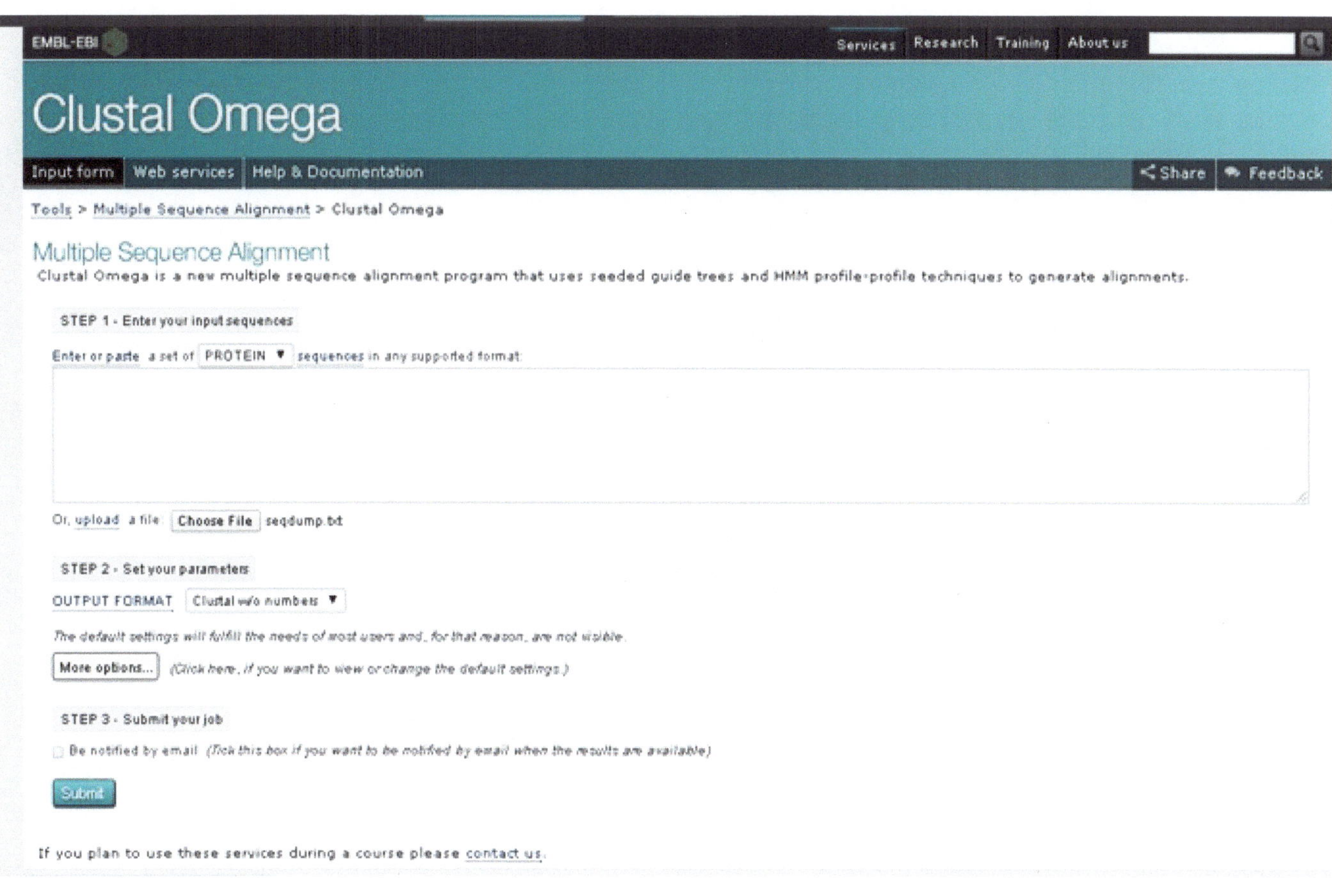
EMBL-EBI
Services Research Training About us
Clustal Omega
Input form Web services Help & Documentation
Share Feedback
Tools > Multiple Sequence Alignment > Clustal Omega
Multiple Sequence Alignment
Clustal Omega is a new multiple sequence alignment program that uses seeded guide trees and HMM profile-profile techniques to generate alignments.
STEP 1 - Enter your input sequences
Enter or paste a set of PROTEIN sequences in any supported format:
Or, upload a file: Choose File seqdump.txt
STEP 2 - Set your parameters
OUTPUT FORMAT Clustal w/o numbers
The default settings will fulfill the needs of most users and, for that reason, are not visible.
More options... (Click here, if you want to view or change the default settings.)
STEP 3 - Submit your job
Be notified by email (Tick this box if you want to be notified by email when the results are available)
Submit
If you plan to use these services during a course please contact us.

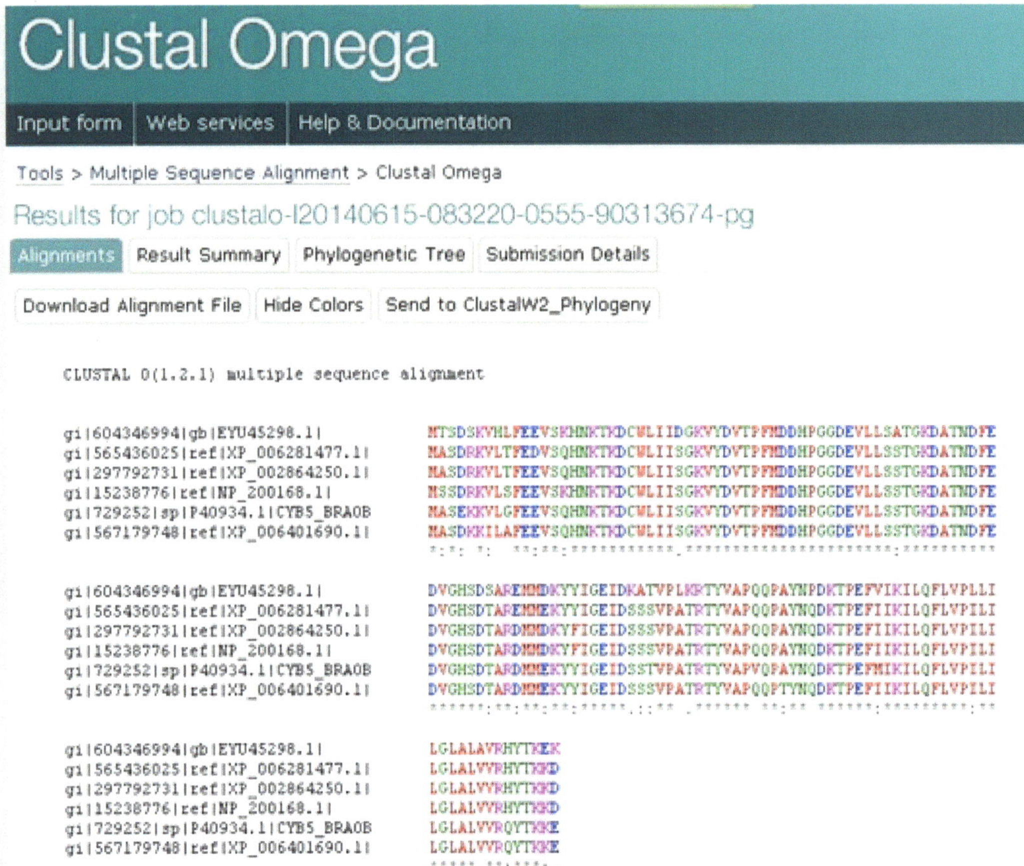

8. EVOLUTIONALRY RELATIONSHIP- PHYLOGENETIC ANALYSIS – TREE VIEW

A **phylogenetic tree** or **evolutionary tree** is a branching diagram or "tree" showing the inferred evolutionary relationships among various biological species or other entities based upon similarities and differences in their physical and/or genetic characteristics. The taxa joined together in the tree are implied to have descended from a common ancestor.

In a **rooted** phylogenetic tree, each node with descendants represents the inferred most recent common ancestor of the descendants, and the edge lengths in some trees may be interpreted as time estimates. Each node is called a taxonomic unit. Internal nodes are generally called hypothetical taxonomic units (HTUs) as they cannot be directly observed. Trees are useful in fields of biology such as bioinformatics, systematics and comparative phylogenetics.

Phylogenetic Tree for given 6 set of sequence such as

gi|604346994|gb|EYU45298.1| 0.12966
gi|565436025|ref|XP_006281477.1| 0.01959
gi|297792731|ref|XP_002864250.1| 0.00622
gi|15238776|ref|NP_200168.1| 0.01617
gi|729252|sp|P40934.1|CYB5_BRAOB 0.03078
gi|567179748|ref|XP_006401690.1| 0.02146

By using CLUSTAL OMEGA- Phylogeny Tool for Phylogram

Phylogenetic Tree

This is a Neighbour-joining tree without distance corrections.

Download Phylogenetic Tree File

```
(
gi|604346994|gb|EYU45298.1|:0.12966,
(
gi|565436025|ref|XP_006281477.1|:0.01959,
(
gi|297792731|ref|XP_002864250.1|:0.00622,
gi|15238776|ref|NP_200168.1|:0.01617)
:0.01026)
:0.00466,
(
gi|729252|sp|P40934.1|CYB5_BRAOB:0.03078,
gi|567179748|ref|XP_006401690.1|:0.02146)
:0.01959);
```

Phylogram

Branch length: Cladogram Real

gi|604346994|gb|EYU45298.1| 0.12966
gi|565436025|ref|XP_006281477.1| 0.01959
gi|297792731|ref|XP_002864250.1| 0.00622
gi|15238776|ref|NP_200168.1| 0.01617
gi|729252|sp|P40934.1|CYB5_BRAOB 0.03078
gi|567179748|ref|XP_006401690.1| 0.02146

By using Tree View Phylogeny Tool for Phylogram

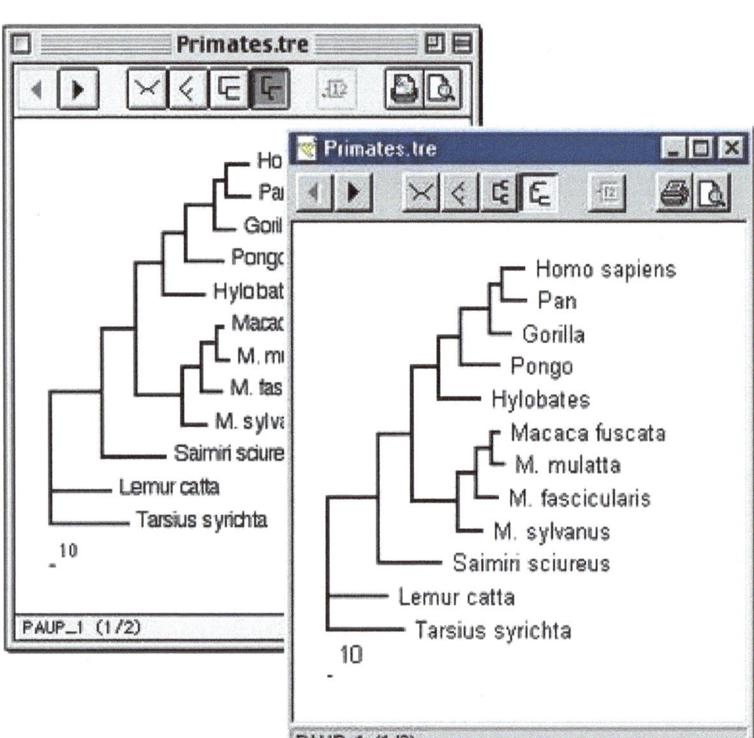

- Download the MSA file from the Clustal OMEGA tool output in .dnd file extension.
- The .dnd file will be downloaded from guided tree in clustal OMEGA output
- To view the graphical of this alignment use Tree view software.
- Download Link for Tree View is http://en.bio-soft.net/tree/TreeView.html
  - Or
    - Google : phylogenetic Treeview Download
    - Select the link such as http://en.bio-soft.net/tree/TreeView.html
    - Do the Setup Process and install this software
- Open the .dnd file in this software.
- It show the close relation among the submitted sequence in three different form.

9. PDB RETRIEVING

- To retrieve the protein structure of any derived protein
- open internet Browser with link http://www.pdb.org/
- In the text search box type query search Eg: Protein name , PDB ID , Organism
- Then submit the search ,The result page will display list of PDB structure with ID
- Click on the interested PDB ID for your requirement and Download the PDB file in PDF format by clicking on the right side Download file .This will retrieve the PDB structure and can visualize in Molecular visualization Tools
    - Such as eg: RASMOL,PYMOL ,MOL MOL

Result : The Given PDB ID 3F3B Retrieved in PDB format.

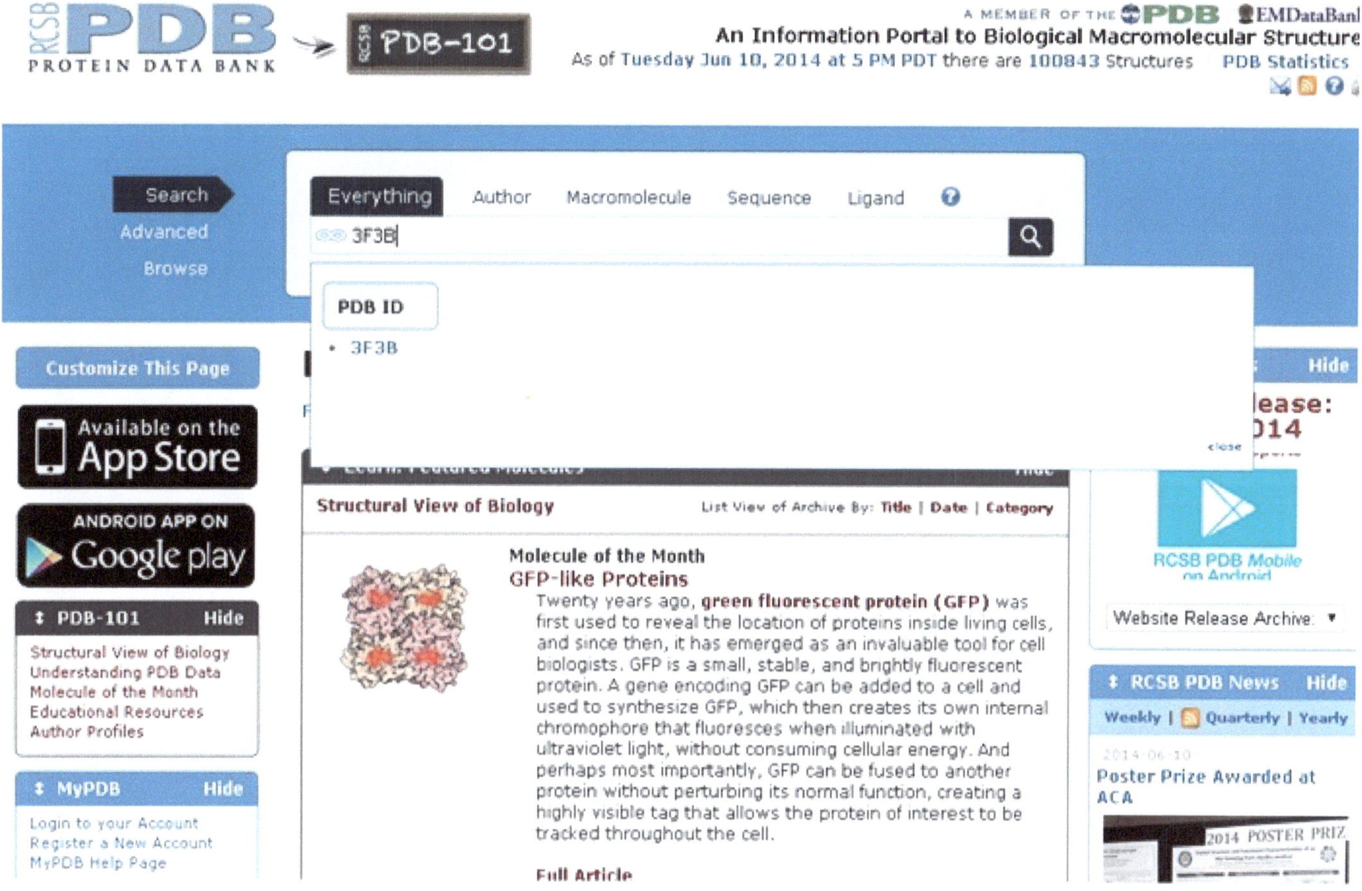

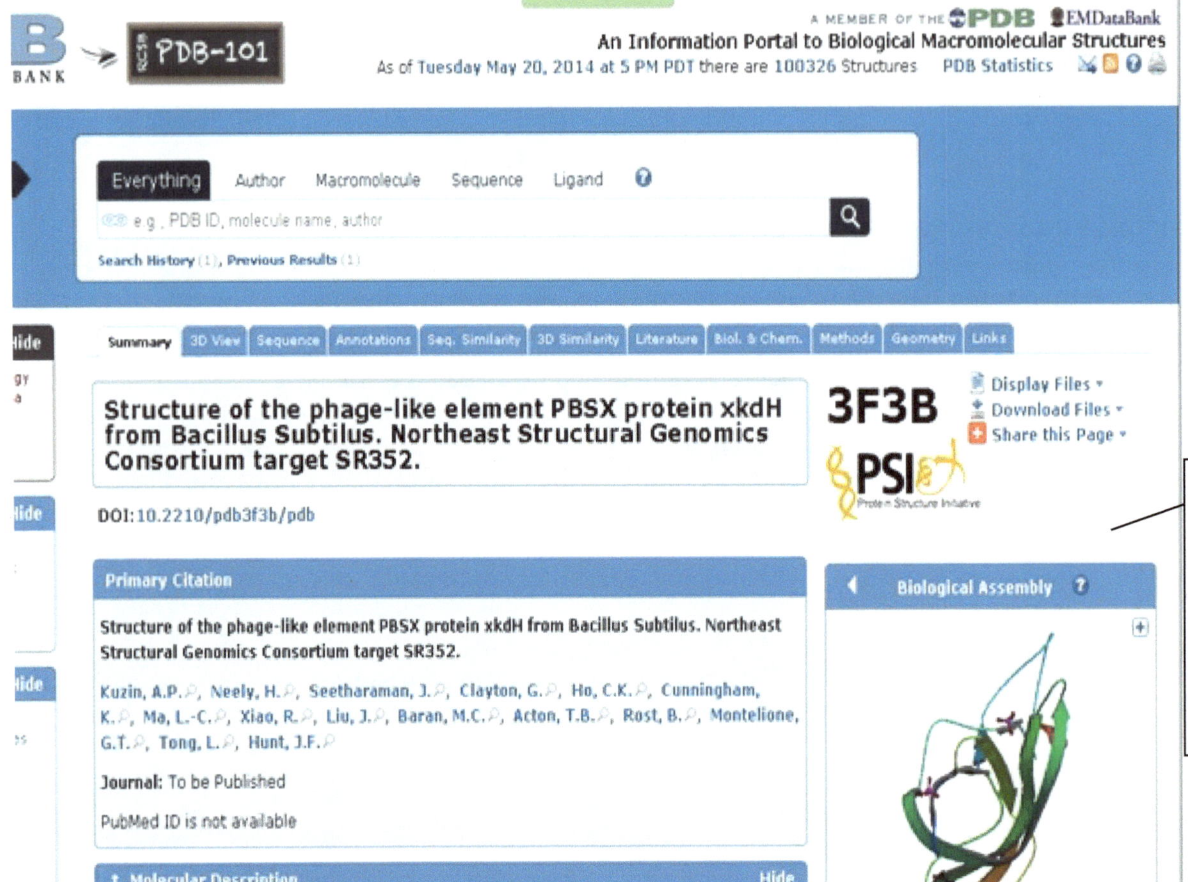

Click here
retrieve pc
file and se
PDB file Te
& save PD
file in your
Folder

11. MOLECULAR VISUALIZATION-RASMOL

- Download the Rasmol Tool from this link http://wbiomed.curtin.edu.au/biochem/help/download.html
- Click on RASWIN ZIP file for Windows operating system and save setup file in your system.
- Unzip or the setup file to install Rasmol in your computer
- Open the Raswin folder click on raswin icon- **RW32B2A** to open rasmol
- This will open two Windows one is for molecule display and other command line window.
- Go to file and Open PDB file this will display the molecule in the Rasmol window.
- Start writing command in Command line Interface i.e sysntax : RasMol> command
- Enter this will display the information of molecule for respective command.

Rasmol Command list for practice.

1.HBonds

Syntax: hbonds {<boolean>}
hbonds <value>

The RasMol hbond command is used to represent the hydrogen bonding of the protein molecule's backbone. This information is useful in assessing the protein's secondary structure. Hydrogen bonds are represented as either dotted lines or cylinders between the donor and acceptor residues. The first time the hbond command is used, the program searches the structure of the molecule to find hydrogen bonded residues and reports the number of bonds to the user. The command hbonds on displays the selected `bonds' as dotted lines, and the hbonds off turns off their display. The colour of hbond objects may be changed by the colour hbond command. Initially, each hydrogen bond has the colours of its connected atoms.

By default the dotted lines are drawn between the accepting oxygen and the donating nitrogen. By using the set hbonds command the alpha carbon positions of the appropriate residues may be used instead. This is especially useful when examining proteins in backbone representation.

2.Show

Syntax: show information
show sequence
show symmetry

The RasMol show command display details of the status of the currently loaded molecule. The command show information lists the molecule's name, classification, PDB code and the number of atoms, chains, groups it contains. If hydrogen bonding, disulphide bridges or secondary structure have been determined, the number of hbonds, ssbonds, helices, ladders and turns are also displayed respectively. The command show sequence lists the residues that compose each chain of the molecule.

Label

Syntax: label {<string>}
label <boolean>

The RasMol label command allows an arbitrary formatted text string to be associated with each currently selected atom. This string may contain embedded `expansion specifiers' which display properties of the atom being labelled. An expansion specifier consists of a `%' character followed by a single alphabetic character specifying the property to be displayed (similar to C's printf syntax). An actual '%' character may be displayed by using the expansion specifier `%%'.

Atom labelling for the currently selected atoms may be turned off with the command label off. By default, if no string is given as a parameter RasMol uses labels appropriate for the current molecule. RasMol uses the label "%n%r:%c.%a" if the molecule contains more than one chain, "%e%i" if the molecule has only a single residue (a small molecule) and "%n%r.%a" otherwise.

The colour of each label may be changed using the colour label command. By default, each label is drawn in the same colour as the atom to which it is attached. The size of the displayed text may be changed using the set fontsize command.

The following table lists the current expansion specifiers:

%a		Atom Name
%b	%t	B-factor/Temperature
%c	%s	Chain Identifier
%e		Element Atomic Symbol
%i		Atom Serial Number
%n		Residue Name
%r		Residue Number

Colour

Syntax: colour {<object>} <colour>
color {<object>} <colour>

Colour the atoms (or other objects) of the selected region. The colour may be given as either a colour name or a comma separated triple of Red, Green and Blue (RGB) components enclosed in square brackets. Typing the command help colours will give a list of all the predefined colour names recognised by RasMol.

Allowed objects are atoms, bonds, backbone, ribbons labels dots, hbonds, and ssbonds. If no object is specified, the default keyword atom is assumed. Some colour schemes are defined for certain object types. The colour scheme none can be applied all objects accept atoms and dots, stating that the selected objects have no colour of their own, but use the colour of their associated atoms (i.e. the atoms they connect). Atom objects can also be coloured by cpk, amino, chain, group, shapely, structure, temperature charge and user. Hydrogen bonds can also be coloured by type and dot surfaces can also be coloured by electrostatic potential. For more information type help colour <colour>.

Set FontSize

Syntax: set fontsize {<value>}

The RasMol set fontsize command is used to control the size of the characters that form atom labels. This value corresponds to the height of the displayed character in pixels. The maximum value of fontsize is 32 pixels, and the default value is 8 pixels high. To display atom labels on the screen use the RasMol label command and to change the colour of displayed labels, use the colour labels command.

SSBonds

Syntax: ssbonds {<boolean>}
ssbonds <value>

The RasMol ssbonds command is used to represent the disulphide bridges of the protein molecule as either dotted lines or cylinders between the connected cysteines. The first time that the ssbonds command is used, the program searches the structure of the protein to find half-cysteine pairs (cysteines whose sulphurs are within 3 angstroms of each other) and reports the number of bridges to the user. The command ssbonds on displays the selected `bonds' as dotted lines, and the command ssbonds off disables the display of ssbonds in the currently selected area. Selection of disulphide bridges is identical to normal bonds, and may be adjusted using the RasMol set bondmode command. The colour of disulphide bonds may be changed using the colour ssbonds command. By default, each disulphide bond has the colours of its connected atoms.
By default disulphide bonds are drawn between the sulphur atoms within the cysteine groups. By using the set ssbonds command the position of the cysteine's alpha carbons may be used instead.

Set

Syntax: set <parameter> {<option>}

The RasMol set command allows the user to alter various internal program parameters such as those controlling rendering options. Each parameter has its own set or permissible parameter options. Typically, ommiting the paramter option resets that parameter to its default value. A list of valid parameter names is given below.

ambient axes background bondmode
boundbox display fontsize hbonds
hetero hourglass hydrogen kinemage
menus mouse radius shadow
slabmode solvent specular specpower
ssbonds strands unitcell vectps

*For more command on Rasmol can refer under downloaded Raswin folder Raswin Help file

Molecular Visualization of PDB -3F3B Structure

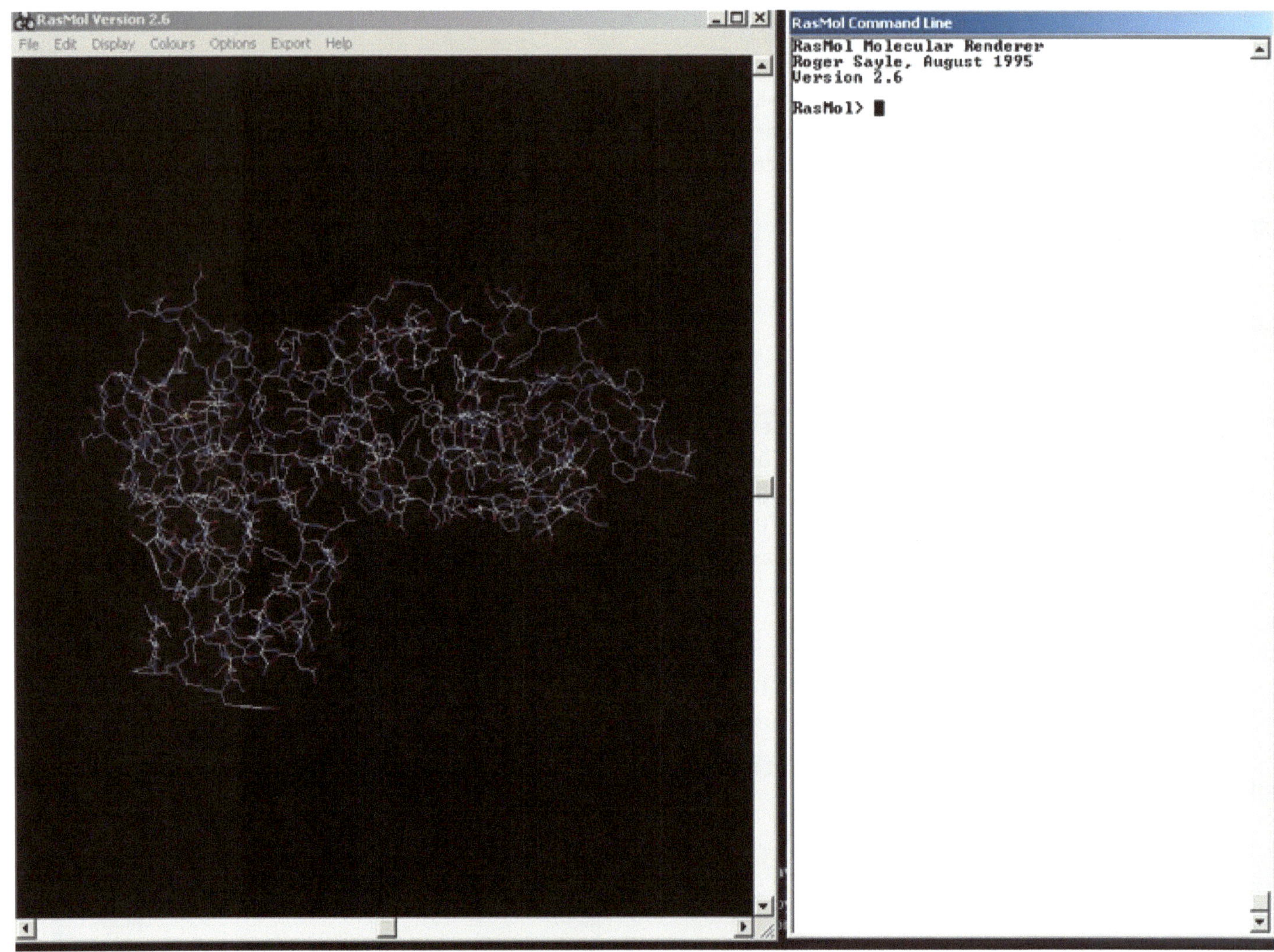

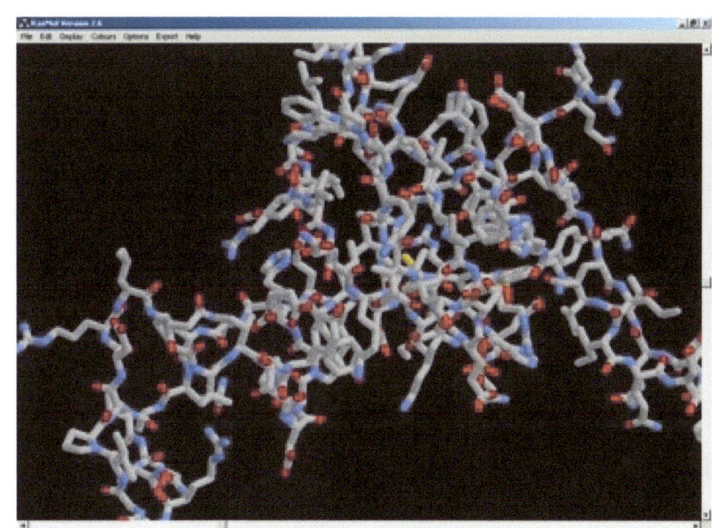

STICK MODEL

SPACE FILL MODEL -WITH LABEL

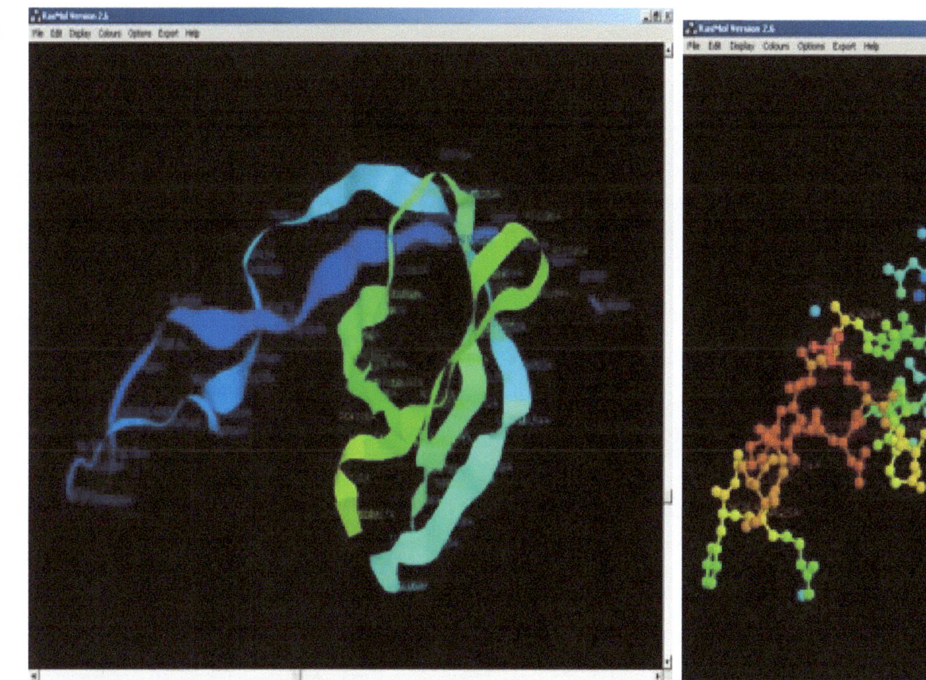

RIBBON MODEL

BALL STICK MODEL

12. IDENTIFICATION OF GENE (HMGA2) IN HUMAN GENOME

- Open internet Browser http ://www.ncbi.nlm.nih.gov.
- Select Gene in search window and give Gene id, Gene name Accession number or name of organism for gene identification
- Click go button for displaying the result sheet.
- The result sheet with different gene information will be displayed.
- Click on the hyperlinked on gene name of your interest.
- The result page will be displayed. Point the mouse on respective GENE it will display information
- Save the file by save as or copy the gene map and save

Result : Identification of **HMGA2** gene in Human Genome is Found at Respective Location.

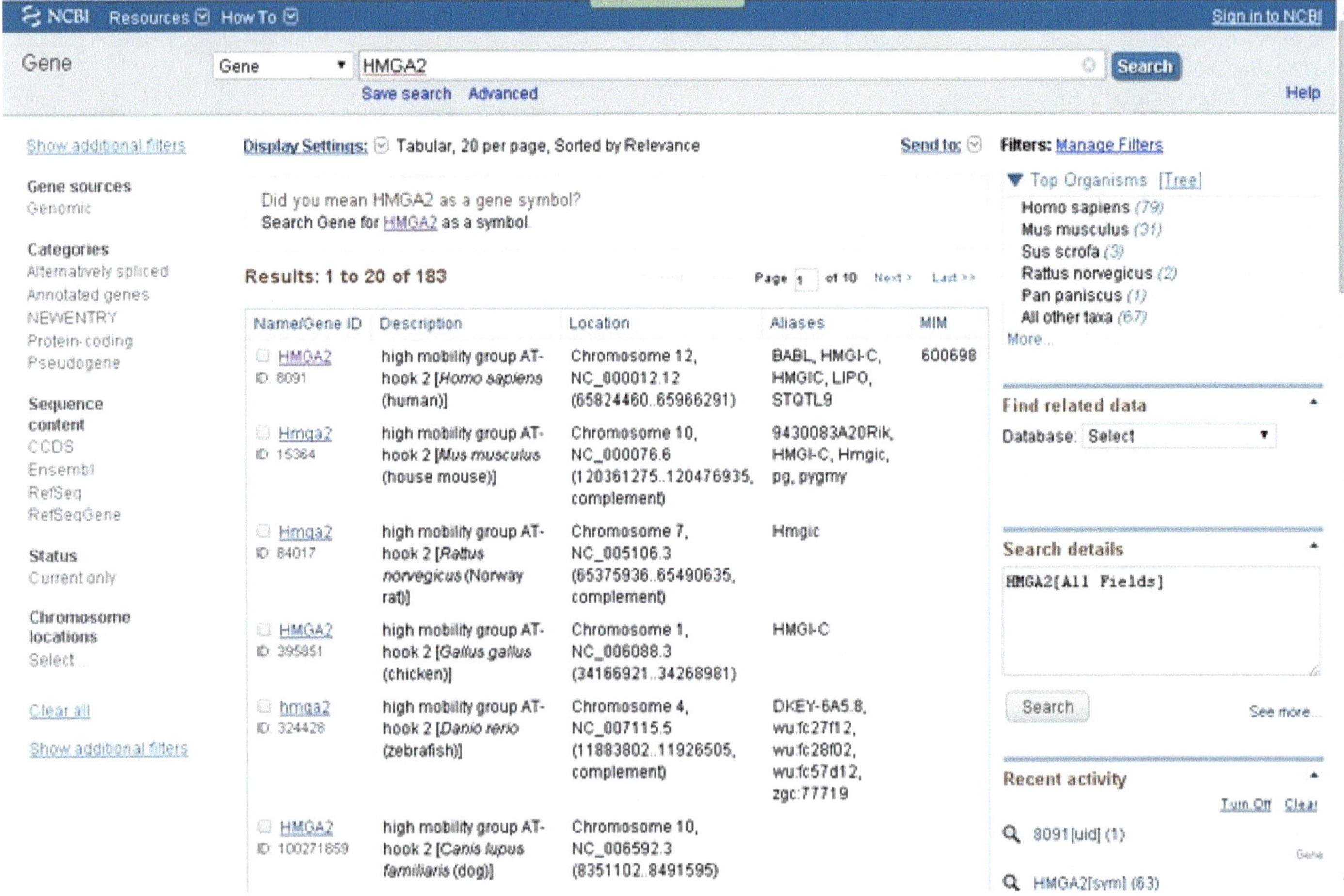

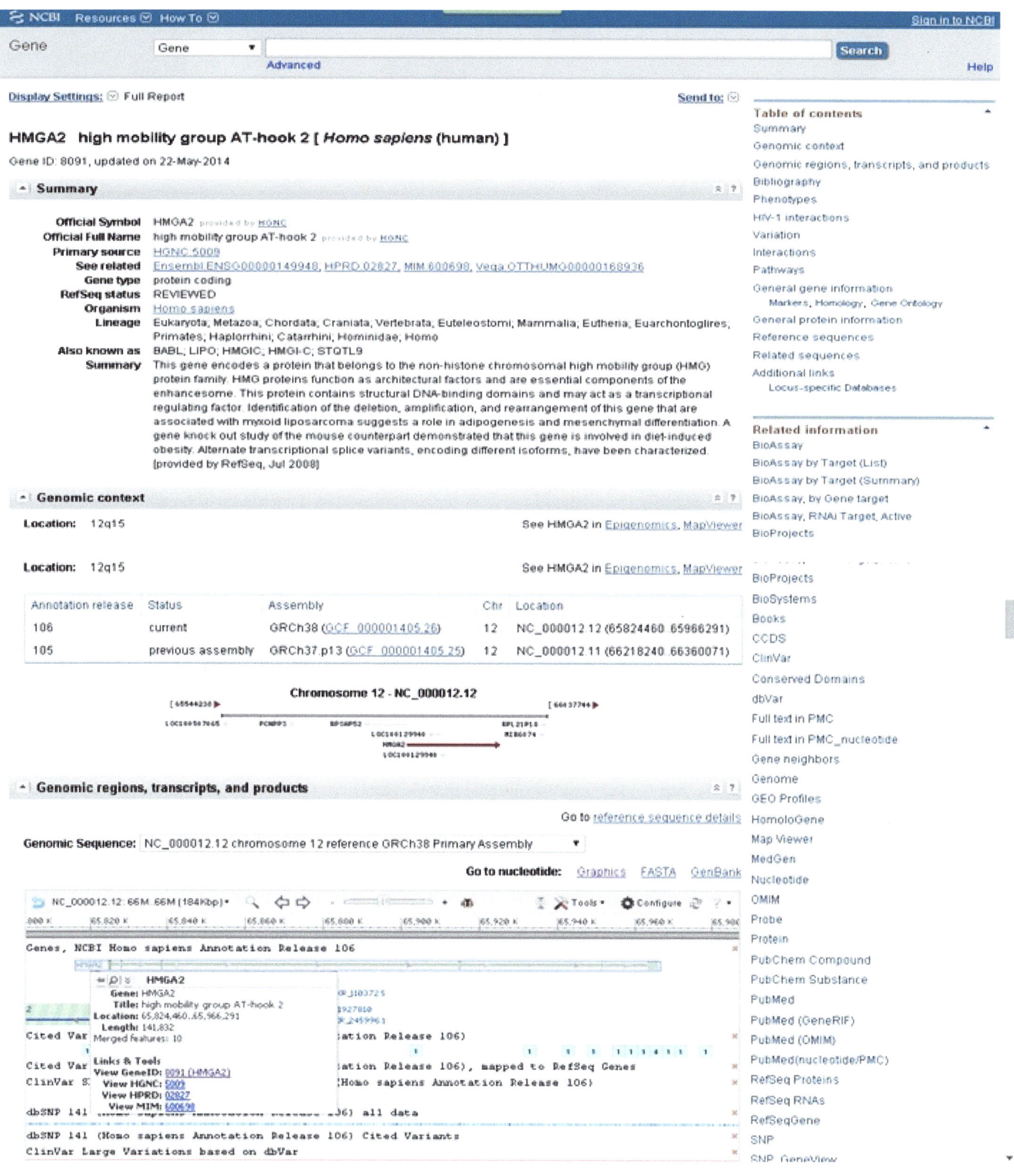
NCBI Resources How To
Sign in to NCBI
Gene
Gene
Advanced
Search
Help
Display Settings: Full Report
Send to:
HMGA2 high mobility group AT-hook 2 [Homo sapiens (human)]
Gene ID: 8091, updated on 22-May-2014
Summary
Official Symbol HMGA2 provided by HGNC
Official Full Name high mobility group AT-hook 2 provided by HGNC
Primary source HGNC:5009
See related Ensembl:ENSG00000149948; HPRD:02827; MIM:600698; Vega:OTTHUMG00000168936
Gene type protein coding
RefSeq status REVIEWED
Organism Homo sapiens
Lineage Eukaryota; Metazoa; Chordata; Craniata; Vertebrata; Euteleostomi; Mammalia; Eutheria; Euarchontoglires; Primates; Haplorrhini; Catarrhini; Hominidae; Homo
Also known as BABL; LIPO; HMGIC; HMGI-C; STQTL9
Summary This gene encodes a protein that belongs to the non-histone chromosomal high mobility group (HMG) protein family. HMG proteins function as architectural factors and are essential components of the enhancesome. This protein contains structural DNA-binding domains and may act as a transcriptional regulating factor. Identification of the deletion, amplification, and rearrangement of this gene that are associated with myxoid liposarcoma suggests a role in adipogenesis and mesenchymal differentiation. A gene knock out study of the mouse counterpart demonstrated that this gene is involved in diet-induced obesity. Alternate transcriptional splice variants, encoding different isoforms, have been characterized. [provided by RefSeq, Jul 2008]
Genomic context
Location: 12q15
See HMGA2 in Epigenomics, MapViewer
Location: 12q15
See HMGA2 in Epigenomics, MapViewer
Annotation release | Status | Assembly | Chr | Location
106 | current | GRCh38 (GCF_000001405.26) | 12 | NC_000012.12 (65824460..65966291)
105 | previous assembly | GRCh37.p13 (GCF_000001405.25) | 12 | NC_000012.11 (66218240..66360071)
Chromosome 12 - NC_000012.12
Genomic regions, transcripts, and products
Go to reference sequence details
Genomic Sequence: NC_000012.12 chromosome 12 reference GRCh38 Primary Assembly
Go to nucleotide: Graphics FASTA GenBank
NC_000012.12: 66M..66M (184Kbp)
Tools
Configure
Genes, NCBI Homo sapiens Annotation Release 106
HMGA2
Gene: HMGA2
Title: high mobility group AT-hook 2
Location: 65,824,460..65,966,291
Length: 141,832
Merged features: 10
Links & Tools
View GeneID: 8091 (HMGA2)
View HGNC: 5009
View HPRD: 02827
View MIM: 600698
Cited Var ation Release 106)
Cited Var ation Release 106), mapped to RefSeq Genes
ClinVar S (Homo sapiens Annotation Release 106)
dbSNP 141 06) all data
dbSNP 141 (Homo sapiens Annotation Release 106) Cited Variants
ClinVar Large Variations based on dbVar
Table of contents
Summary
Genomic context
Genomic regions, transcripts, and products
Bibliography
Phenotypes
HIV-1 interactions
Variation
Interactions
Pathways
General gene information
Markers, Homology, Gene Ontology
General protein information
Reference sequences
Related sequences
Additional links
Locus-specific Databases
Related information
BioAssay
BioAssay by Target (List)
BioAssay by Target (Summary)
BioAssay, by Gene target
BioAssay, RNAi Target, Active
BioProjects
BioProjects
BioSystems
Books
CCDS
ClinVar
Conserved Domains
dbVar
Full text in PMC
Full text in PMC_nucleotide
Gene neighbors
Genome
GEO Profiles
HomoloGene
Map Viewer
MedGen
Nucleotide
OMIM
Probe
Protein
PubChem Compound
PubChem Substance
PubMed
PubMed (GeneRIF)
PubMed (OMIM)
PubMed(nucleotide/PMC)
RefSeq Proteins
RefSeq RNAs
RefSeqGene
SNP

Primer Designing

A primer is a short synthetic oligonucleotide which is used in many molecular techniques from PCR to DNA sequencing. These primers are designed to have a sequence which is the reverse complement of a region of template or target DNA to which we wish the primer to anneal.

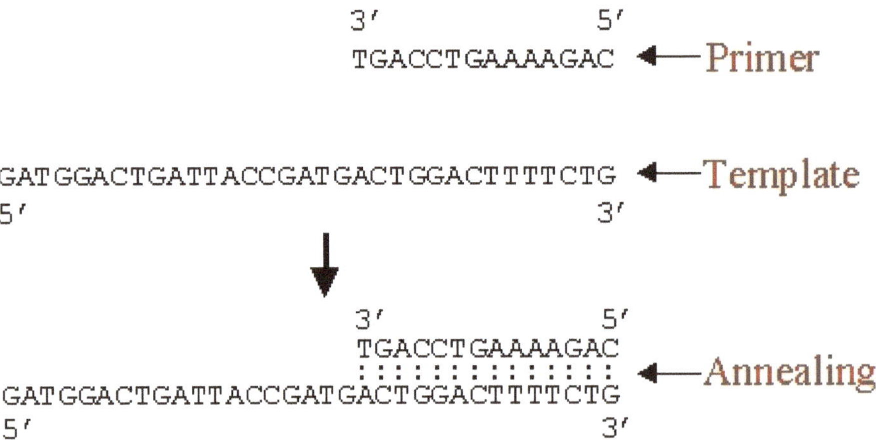

Some thoughts on designing primers.

1. primers should be 17-28 bases in length;
2. base composition should be 50-60% (G+C);
3. primers should end (3') in a G or C, or CG or GC: this prevents "breathing" of ends and increases efficiency of priming;
4. Tms between 55-80°C are preferred;
5. 3'-ends of primers should not be complementary (ie. base pair), as otherwise primer dimers will be synthesised preferentially to any other product;
6. primer self-complementarity (ability to form 2° structures such as hairpins) should be avoided;
7. runs of three or more Cs or Gs at the 3'-ends of primers may promote mispriming at G or C-rich sequences (because of stability of annealing), and should be avoided.

PRIMER PICKING FORM RETRIVED SEQUENCE

- Open any Internet Browser
- Go to the link http://www.ncbi.nlm.nih.gov
- On the top of NCBI home page drag down All database and Select Protein.
- In the search text box type key word of query such as Accession number (or) disease name (or) organism name to retrieve the sequence.
- Click on Search button this will display list of available result match to the given keyword.
- Select the check box in the result page which you want to retrieve the sequence.
- Go to display setting and select FASTA format .Because most of the bioinformatics tools accept FASTA file format.
- The result will display in FASTA format

- Save the file by save as in the top right corner by clicking button send file
- Or Copy the file text from '>' (greater symbol) up to the End of the sequence text and save in notepad with file name and extension as **.FASTA .**
- In the FASTS result home page click the link pick primer in the Left OR open Primer Tool
- With this link http://www.ncbi.nlm.nih.gov/tools/primer-blast/ and paste your FASTA Sequence or Gi ID
- In the Primer Tool home page check with this condition in the Range for Forward and Reverse option
- Enter the position ranges if you want the primers to be located on the specific sites. The positions refer to the base numbers on the plus strand of your template (i.e., the "From" position should always be smaller than the "To" position for a given primer). Partial ranges are allowed. For example, if you want the PCR product to be located between position 100 and position 1000 on the template, you can set forward primer "From" to 100 and reverse primer "To" to 1000 (but leave the forward primer "To" and reverse primer "From" empty).
  Note that the position range of forward primer may not overlap with that of reverse primer.
- Tick on Primer Pair Specificity Checking Parameters
- Click on get primer button

The selected Sequence GI Id is 498488879 for Picking PRIMERS

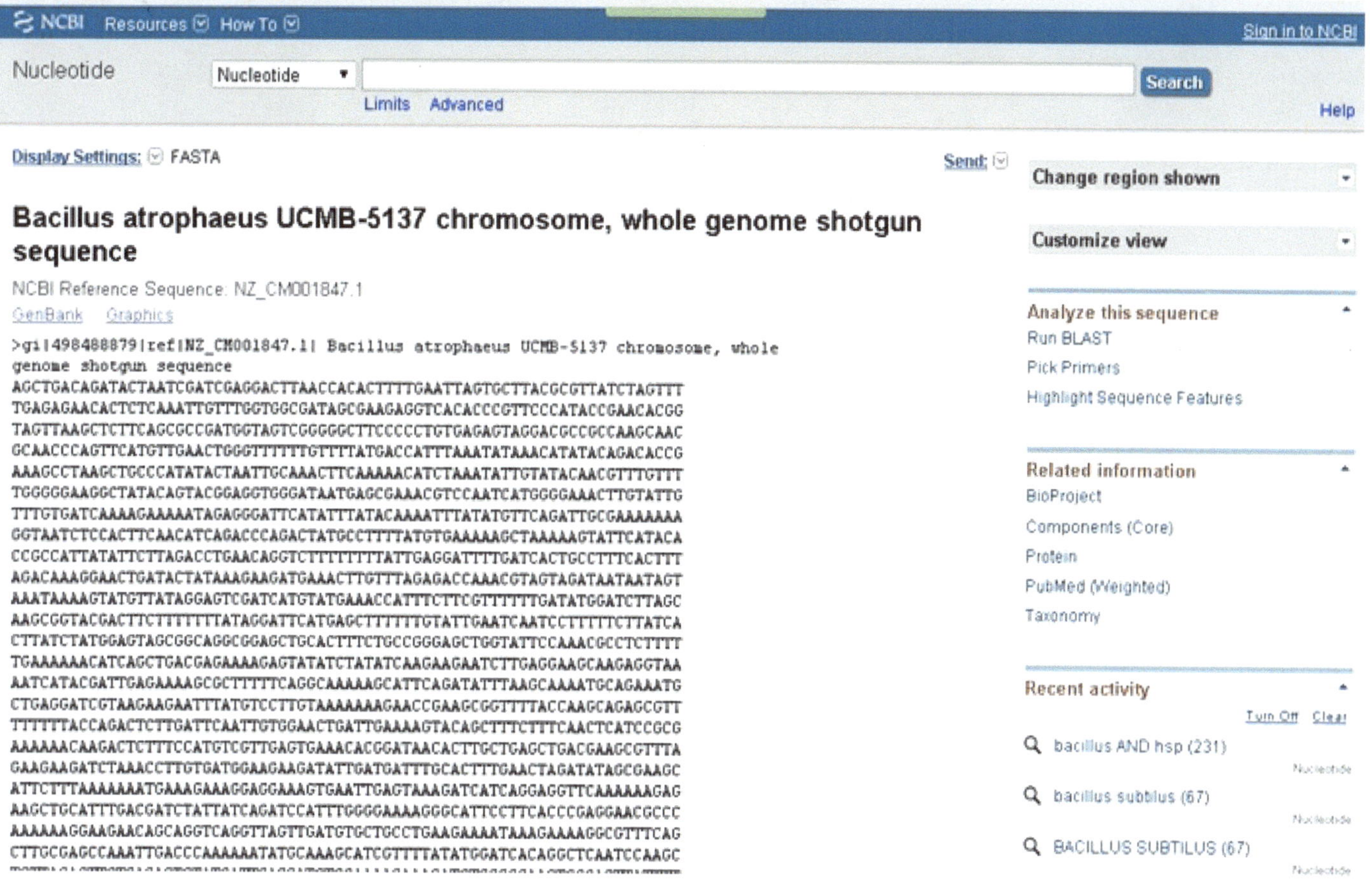

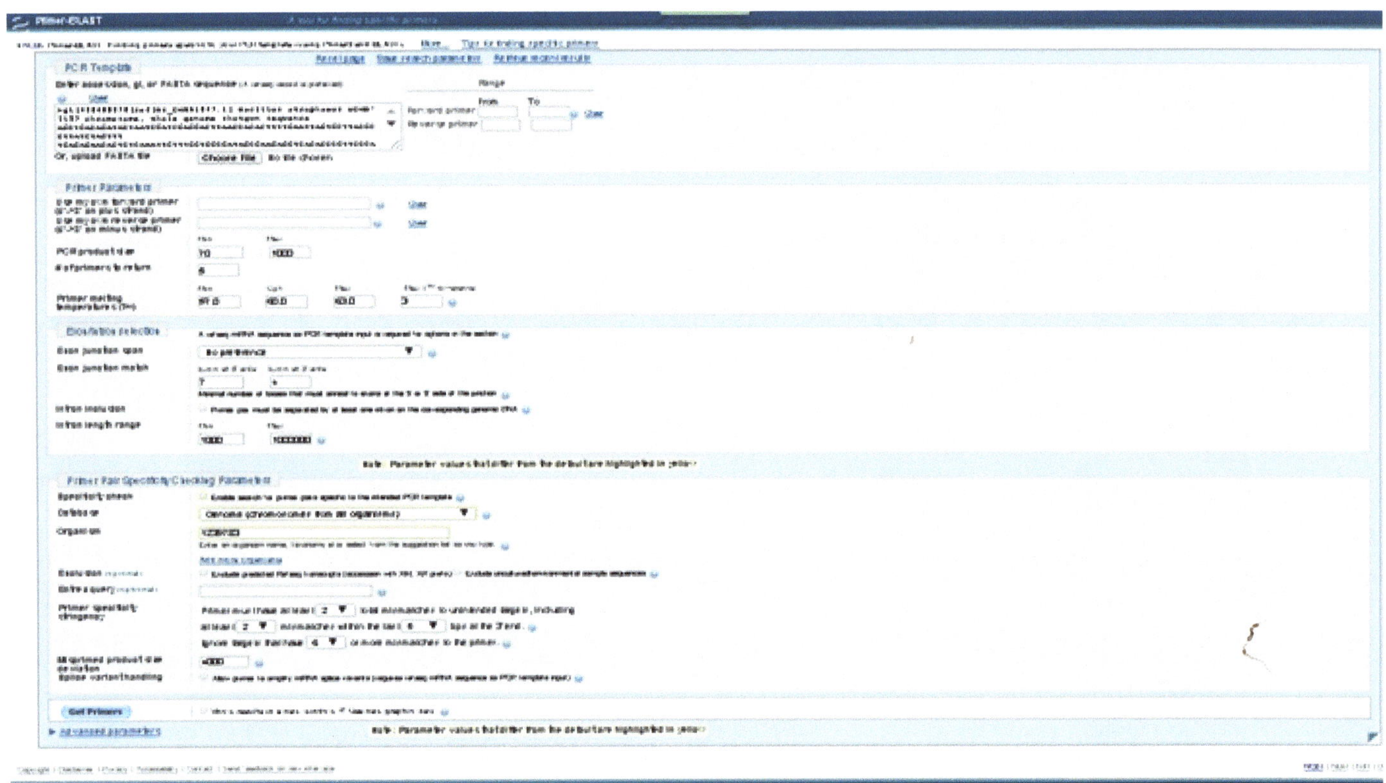

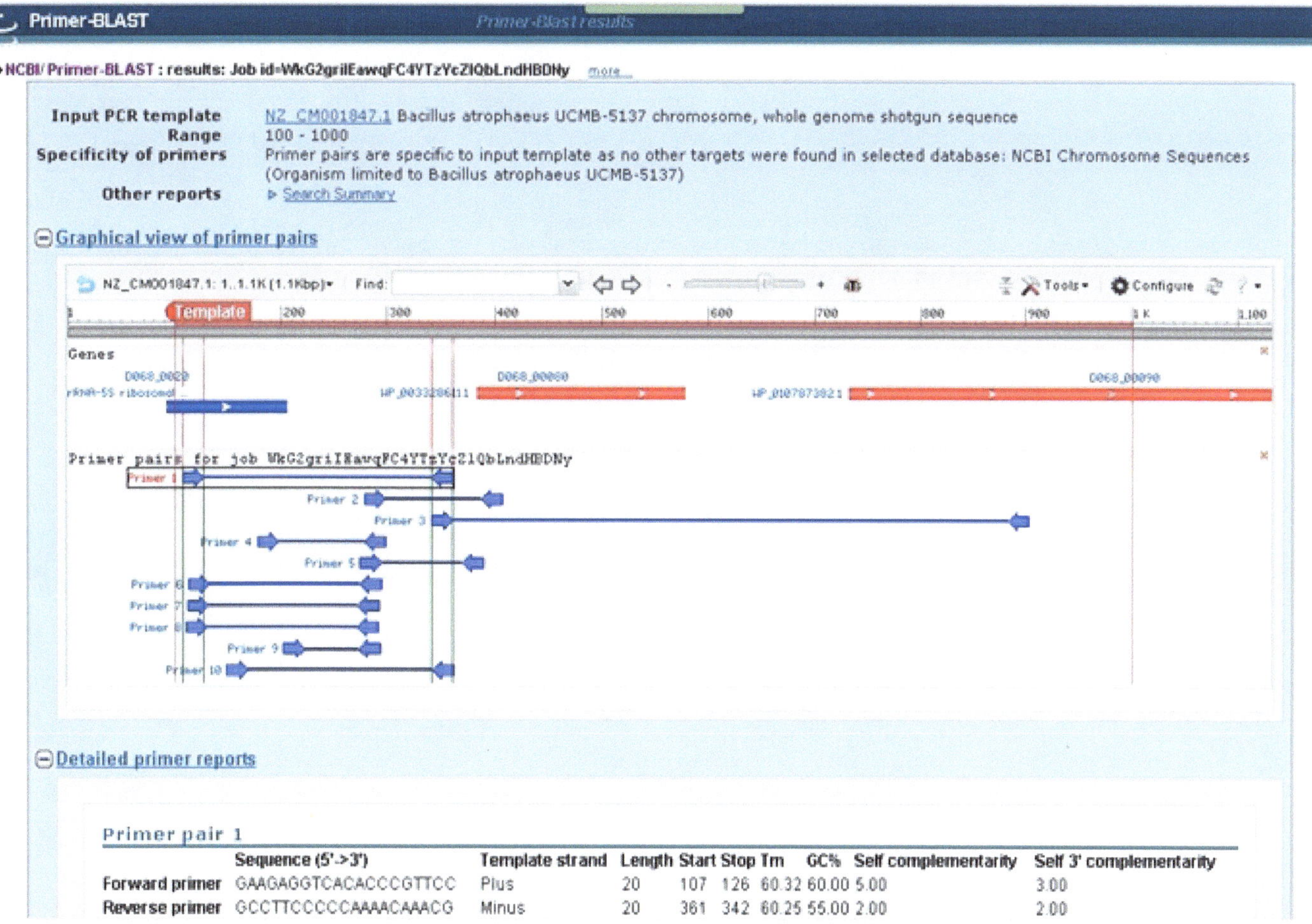
Primer-BLAST
Primer-Blast results
NCBI/ Primer-BLAST : results: Job id=WkG2griIEawqFC4YTzYcZlQbLndHBDNy more...
Input PCR template NZ_CM001847.1 Bacillus atrophaeus UCMB-5137 chromosome, whole genome shotgun sequence
Range 100 - 1000
Specificity of primers Primer pairs are specific to input template as no other targets were found in selected database: NCBI Chromosome Sequences (Organism limited to Bacillus atrophaeus UCMB-5137)
Other reports Search Summary
Graphical view of primer pairs
NZ_CM001847.1: 1..1.1K (1.1Kbp)
Find:
Tools
Configure
Template
Genes
Primer pairs for job WkG2griIEawqFC4YTzYcZlQbLndHBDNy
Primer 1
Primer 2
Primer 3
Primer 4
Primer 5
Primer 6
Primer 7
Primer 8
Primer 9
Primer 10
Detailed primer reports
Primer pair 1
Sequence (5'->3') Template strand Length Start Stop Tm GC% Self complementarity Self 3' complementarity
Forward primer GAAGAGGTCACACCCGTTCC Plus 20 107 126 60.32 60.00 5.00 3.00
Reverse primer GCCTTCCCCCAAAACAAACG Minus 20 361 342 60.25 55.00 2.00 2.00

Primer-BLAST Primer-Blast results

▸NCBI/ Primer-BLAST : results: Job id=WkG2grilEawqFC4YTzYcZlQbLndHBDNy more...

Input PCR template	NZ_CM001847.1 Bacillus atrophaeus UCMB-5137 chromosome, whole genome shotgun sequence
Range	100 - 1000
Specificity of primers	Primer pairs are specific to input template as no other targets were found in selected database: NCBI Chromosome Sequences (Organism limited to Bacillus atrophaeus UCMB-5137)
Other reports	▸ Search Summary

⊕Graphical view of primer pairs

⊖Detailed primer reports

Primer pair 1

	Sequence (5'->3')	Template strand	Length	Start	Stop	Tm	GC%	Self complementarity	Self 3' complementarity
Forward primer	GAAGAGGTCACACCCGTTCC	Plus	20	107	126	60.32	60.00	5.00	3.00
Reverse primer	GCCTTCCCCCAAAACAAACG	Minus	20	361	342	60.25	55.00	2.00	2.00
Product length	255								

Products on intended target

▸NZ_CM001847.1 Bacillus atrophaeus UCMB-5137 chromosome, whole genome shotgun sequence

```
product length = 255
Features associated with this product:
   rRNA-5S ribosomal RNA

Forward primer  1     GAAGAGGTCACACCCGTTCC  20
Template        107   ....................  126

Reverse primer  1     GCCTTCCCCCAAAACAAACG  20
Template        361   ....................  342
```

Primer pair 2

	Sequence (5'->3')	Template strand	Length	Start	Stop	Tm	GC%	Self complementarity	Self 3' complementarity
Forward primer	CGAAAGCCTAAGCTGCCCAT	Plus	20	279	298	60.75	55.00	6.00	2.00
Reverse primer	CCCATGATTGGACGTTTCGC	Minus	20	407	388	59.90	55.00	5.00	2.00
Product length	129								

12. Introduction to AutoDock

AutoDock is a suite of automated docking tools. It is designed to predict how small molecules, such as substrates or drug candidates, bind to a receptor of known 3D structure.

Current distributions of AutoDock consist of two generations of software: AutoDock 4 and AutoDock Vina.

AutoDock 4 actually consists of two main programs: *autodock* performs the docking of the ligand to a set of grids describing the target protein; *autogrid* pre-calculates these grids.

In addition to using them for docking, the atomic affinity grids can be visualised. This can help, for example, to guide organic synthetic chemists design better binders.

AutoDock Vina does not require choosing atom types and pre-calculating grid maps for them. Instead, it calculates the grids internally, for the atom types that are needed, and it does this virtually instantly.

We have also developed a graphical user interface called AutoDockTools, or ADT for short, which amongst other things helps to set up which bonds will treated as rotatable in the ligand and to analyze dockings.

AutoDock has applications in:

- X-ray crystallography;
- structure-based drug design;
- lead optimization;
- virtual screening (HTS);
- combinatorial library design;
- protein-protein docking;
- chemical mechanism studies.

AutoDock 4 is free and is available under the GNU General Public License. AutoDock Vina is available under the Apache license, allowing commercial and non-commercial use and redistribution. Click on the "Downloads" tab. And Happy Docking!

What is AutoDock Vina?

AutoDock Vina is a new generation of docking software from the Molecular Graphics Lab. It achieves significant improvements in the average accuracy of the binding mode predictions, while also being up to two orders of magnitude faster than AutoDock 4.[1]

Because the scoring functions used by AutoDock 4 and AutoDock Vina are different and inexact, on any given problem, either program may provide a better result.

AutoDock 4.2 is faster than earlier versions, and it allows sidechains in the macromolecule to be flexible. As before, rigid docking is blindingly fast, and high-quality flexible docking can be done in around a minute. Up to 40,000 rigid dockings can be done in a day on one cpu.

AutoDock 4.2 now has a free-energy scoring function that is based on a linear regression analysis, the AMBER force field, and an even larger set of diverse protein-ligand complexes with known inhibition constants than we used in AutoDock 3.0. The best model was cross-validated with a separate set of HIV-1 protease complexes, and confirmed that the standard error is around 2.5 kcal/mol. This is enough to discriminate between leads with milli-, micro- and nano-molar inhibition constants.

You can read more about the new features in AutoDock 4.2 and how to use them in the AutoDock4.2 User Guide.

The introduction of AutoDock 4 comprises three major improvements:

1. The docking results are more accurate and reliable.
2. It can optionally model flexibility in the target macromolecule.
3. It enables AutoDock's use in evaluating protein-protein interactions.

AutoDock 4.0 not only is it faster than earlier versions, it allows sidechains in the macromolecule to be flexible. As before, rigid docking is blindingly fast, and high-quality flexible docking can be done in around a minute. Up to 40,000 rigid dockings can be done in a day on one cpu.

AutoDock 4.0 now has a free-energy scoring function that is based on a linear regression analysis, the AMBER force field, and an even larger set of diverse protein-ligand complexes with known inhibiton constants than we used in AutoDock 3.0. The best model was cross-validated with a separate set of HIV-1 protease complexes, and confirmed that the standard error is around 2.5 kcal/mol. This is enough to discriminate between leads with milli-, micro- and nano-molar inhibition constants.

.

AutoDock 4.0 can be compiled to take advantiage of new search methods from the optimization library, ACRO, developed by William E. Hart at Sandia National Labs. We have also added some new features to our existing evolutionary methods. We still provide the Monte Carlo simulated annealing (SA) method of 2.4 and earlier. The Lamarckian Genetic Algorithm (LGA) is a big improvement on the Genetic Algorithm, and both genetic methods are much more efficient and robust than SA.

What is AutoDockTools (ADT)?

We have developed and continue to improve our graphical front-end for AutoDock and AutoGrid, ADT (AutoDockTools). It runs on Linux, Mac OS X, SGI IRIX and Microsoft Windows. We also have new tutorials, along with accompanying sample files.

Where is AutoDock Used?

AutoDock has now been distributed to more than 29000 users around the world. It is being used in academic, governmental, non-profit and commercial settings. In January of 2011, a search of the ISI Citation Index showed more than 2700 publications have cited the primary AutoDock methods papers.

AutoDock is now distributed under the GPL open source license and is freely available for all to use. Because of the restrictions of incorporating GPL licensed software into other codes for the purpose of redistribution, some companies may wish to license AutoDock under a separate license agreement - which we can arrange. Please contact Prof. Arthur J. Olson at + 1 (858) 784-2526 for more information.

Why Use AutoDock?

AutoDock has been widely-used and there are many examples of its successful application in the literature (see References); in 2006, AutoDock was the most cited docking software. It is very fast, provides high quality predictions of ligand conformations, and good correlations between predicted inhibition constants and experimental ones. AutoDock has also been shown to be useful in blind docking, where the location of the binding site is not known. Plus, AutoDock is free software and version 4 is distributed under the GNU General Public License

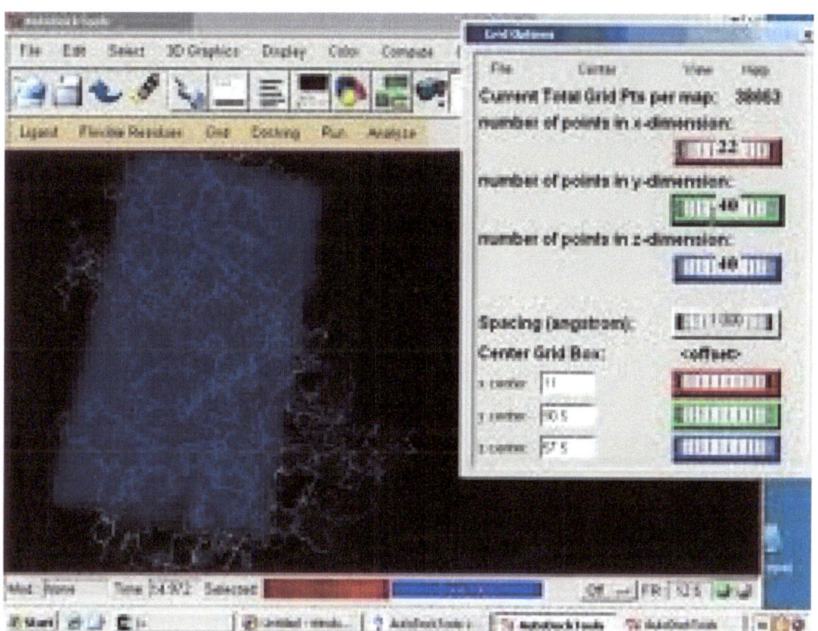

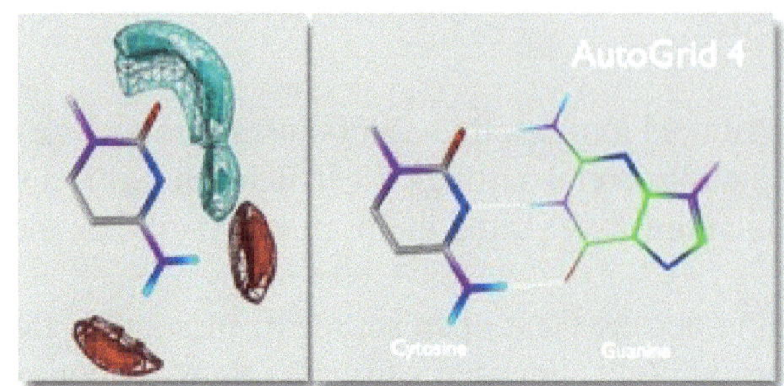
AutoGrid 4
Cytosine
Guanine

13. CALCULATION OF STANDARD DEVIATION, VARIANCE, GRAPH PLOT – MS EXCEL

STANDARD DEVIATION

- In windows XP operating system
- Go to start→ program→Ms office→Ms excel
- Get the value of samples such as x-value ,y-value.
- Put the x value in one column egg: column A
  Put the y value in one column egg: column B
- Note the column name: number for which the SD should calculate egg: B2 :B9.
- Place the cursor in new column to write the formula.
  Write the Standard deviation MS –excel formula i.e SD [=STDEV(B2:B9)]
- Run the formula for respective calculation.

VARIANCE

- In windows XP operating system
- Go to start→ program→Ms office→Ms excel
- Get the value of samples such as x-value ,y-value.
- Put the x value in one column egg: column A
  Put the y value in one column egg: column B
- Note the column name: number for which the variance should calculate egg: B2 :B9.
- Place the cursor in new column to write the formula.
  Write the variance MS –excel formula i.e [=VAR(B2:B9)]
- Run the formula for respective calculation

GRAPH PLOTTING

- In windows XP operating system
- Go to start→ program→Ms office→Ms excel
- Get the value of samples such as x-value ,y-value.
- Put the x value in one column egg: column A
- Put the y value in one column egg: column B
- Select both the column values from the excel sheet.
- Go to insert, in the tool menu .Select other charts or column based on your requirement of graph to plot.
- Select any one type of chart to plot the graph & click ok.
- Then the selected values graph is displayed.

14. OPEN READING FRAME –ORF

- Open internet browser with link http://www.ncbi.nlm.nih.gov
- Retrieve the nucleotide sequence in FASTA format and save it.
- Open ORF tool from NCBI home page link : http://www.ncbi.nlm.nih.gov/gorf/gorf.html
- Paste the GI or ACCESSION number in the given box (or) paste sequence in FASTA format in given text box provided
- Click on Orf Find button
- The result will display all possible ORF's in 6 frames
- Click on longest ORF. This will give number of amino acid residues
- The result sequence will have the start codon and the stop codon

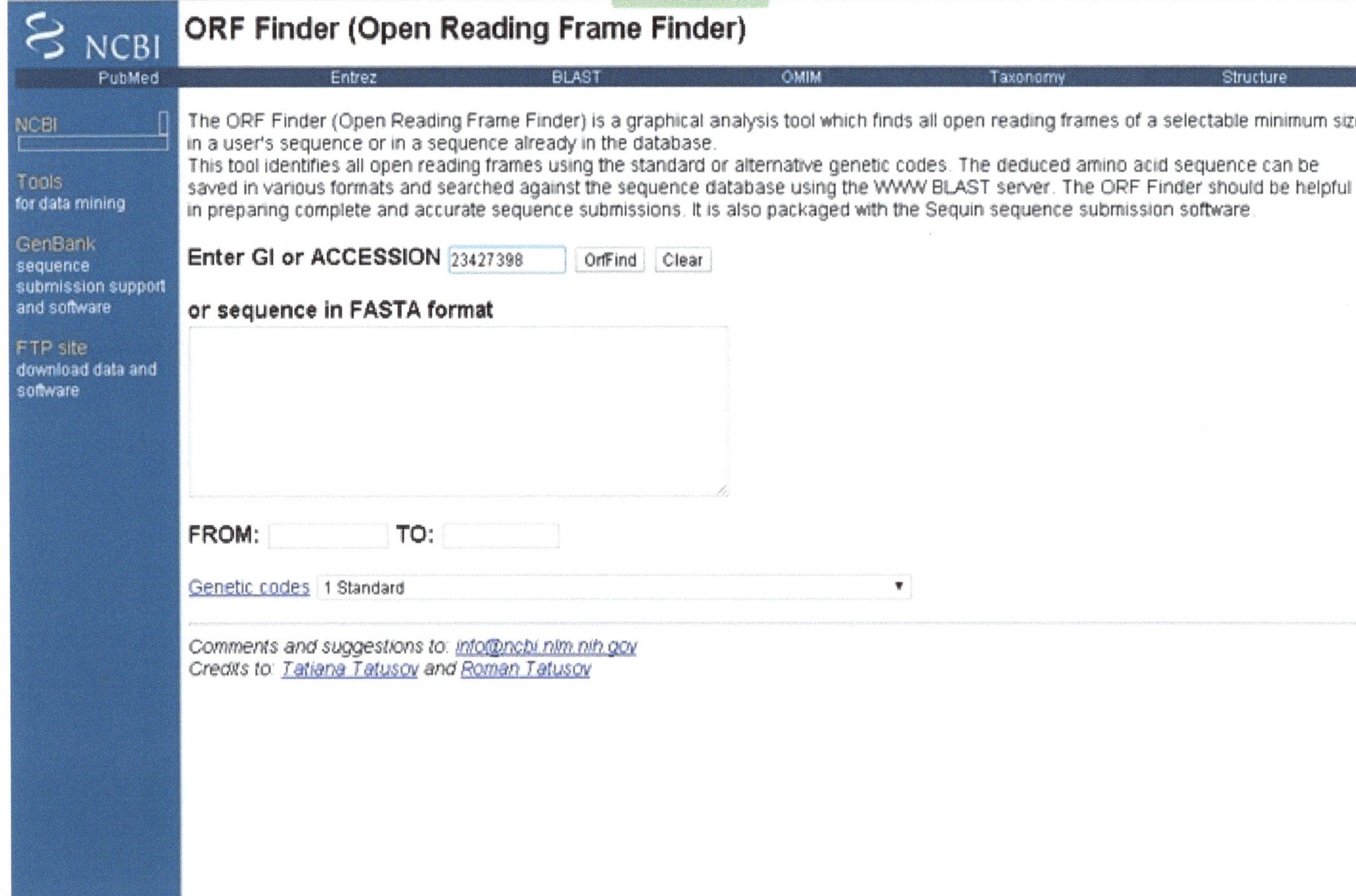

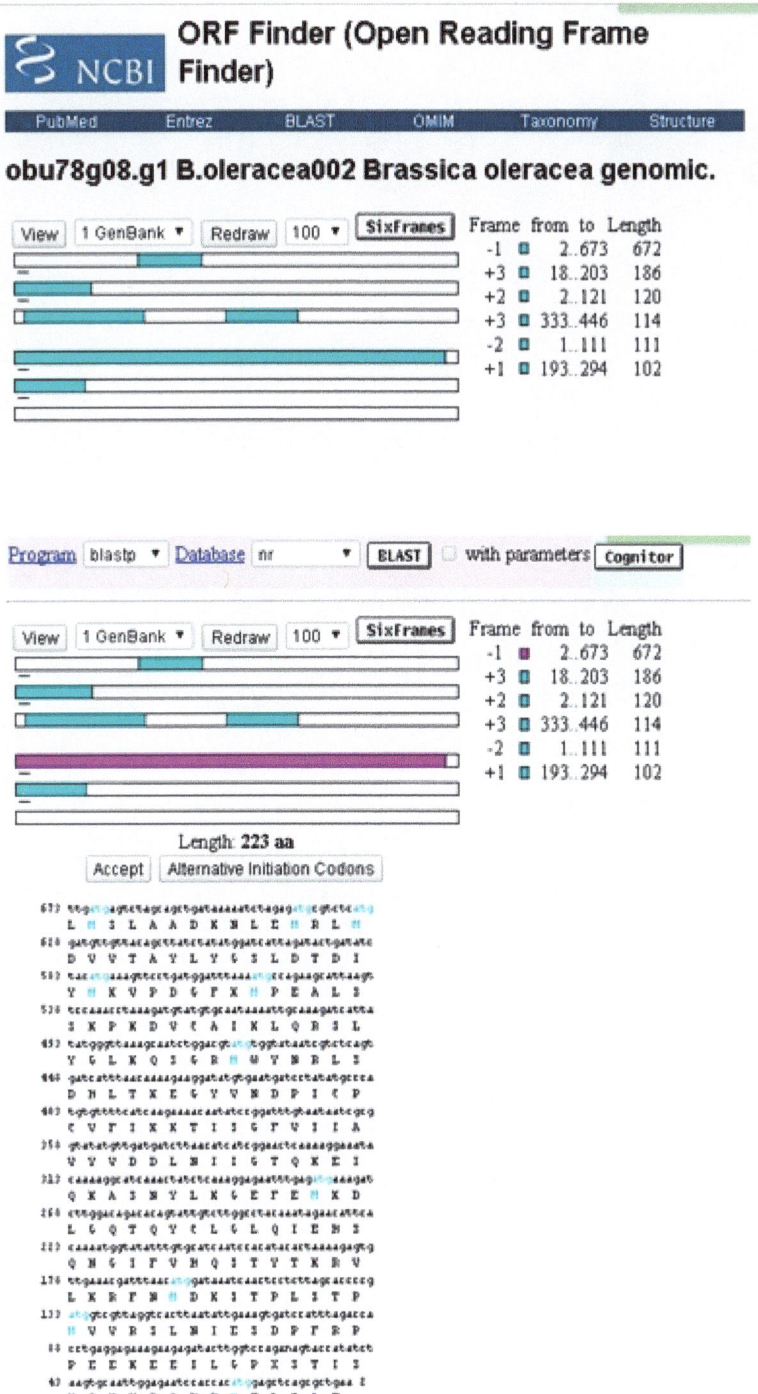
ORF Finder (Open Reading Frame Finder)
NCBI
PubMed Entrez BLAST OMIM Taxonomy Structure
obu78g08.g1 B.oleracea002 Brassica oleracea genomic.
View 1 GenBank Redraw 100 SixFrames
Frame from to Length
-1 2..673 672
+3 18..203 186
+2 2..121 120
+3 333..446 114
-2 1..111 111
+1 193..294 102
Program blastp Database nr BLAST with parameters Cognitor
View 1 GenBank Redraw 100 SixFrames
Frame from to Length
-1 2..673 672
+3 18..203 186
+2 2..121 120
+3 333..446 114
-2 1..111 111
+1 193..294 102
Length: 223 aa
Accept Alternative Initiation Codons

Ramachandran Plot

The Ramachandran Plot Window plots values for the presently chose amino-acids of the current layer. The name of the current layer is drawn at the bottom left of the window. Amino-acids show up as a bit cross except for Gly that shows up as a square. To know to which amino-corrosive a point has a place, essentially put the mouse pointer onto a point or a gathering of focuses. Its (their) name(s) will show up at the highest point of the Window.

Omega, Phi, Psi and optional structure work of the presently showed focuses could be spared as a classified content record for further dissection (Cmd + s on a Macintosh, Ctrl+s on a PC). In addition, a picture can additionally be sent out (Macintosh just). Center and permitted areas as characterized by Morris et al. are likewise surrounded, and it is conceivable to select amino-acids that are outside of these locales in one operation with the suitable thing of the select menu.

Note: The Ramachandran plot could be utilized to specifically alter a Phi/Psi edge of a buildup with an immediate reaction of the change. Everything you need to do is drag a point at its new area. You can compel the pivot around either of the Phi or Psi hub by holding down the "9" or "0" key while moving the point. Naturally it is the C-terminal some piece of the protein that will move, as a bit "c" drawn simply underneath the help symbol reminds you. On the off chance that you need the N-terminal part to move, click on the "c" that will be changed into a 'n'. In the event that you need to move just a piece of the spine (and not the entire spine up to the C-terminal), you must break the spine between the last amino-corrosive you need to move and the first you need to stay altered. This is finished with the proper thing of the instrument menu.

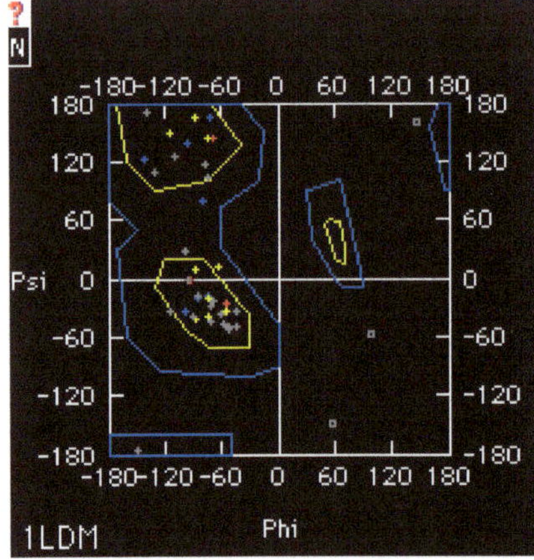

www.ingramcontent.com/pod-product-compliance
Lightning Source LLC
Chambersburg PA
CBHW050808180526
45159CB00004B/1594

* 9 7 8 1 5 1 7 5 0 5 8 8 2 *